Traditional Witches' History of the Occult Banking System

How Witches and Occultists Can Use Bitcoin and Altcoins for Privacy and Anti-discrimination

Sophia diGregorio
2018

Winter Tempest
Books

Traditional Witches' History of the Occult Banking System

How Witches and Occultists Can Use Bitcoin and Altcoins for Privacy and Anti-discrimination

Sophia diGregorio

2018

Winter Tempest Books

DEDICATION

I dedicate this book to you, the reader. Although history and finance are often considered drab subjects, I hope you enjoy reading this book as much as I enjoyed writing it and I hope it provides you with new ways of thinking about and doing ordinary things. Above all, I hope this book makes your life better.

CONTENTS

PREFACE

The official history is written by our persecutors and, as such, represents only a distorted fragment of our past. Their account of events mostly ignores our existence, however, when we are included it is as consorts of the devil, temptresses, tormentors of religious men, soulless animals, enemies of God, and the embodiment of Satan. We are the witches, the living descendants of those accused, tortured, murdered, and silenced. While we have survived into the twenty-first century, so have our worst enemies who have now spent thousands of years erecting a nearly worldwide pyramid of domination with themselves perched on top.

In Europe, North America, and everywhere else they invaded and colonized, these dark overlords looted genuine things and replaced them with their own fraudulent versions, whether it was religion, medicine, science, or money, to create a network of systems constituting a monolithic, virtual prison. They perpetrated their deception from one generation to the next, evermore narrowing humanity's perceptions, while styling an elaborate, limited fictional world, which, to most people, appears perfectly natural.

They easily accomplished this because, for centuries, they fully controlled the flow of information. Theirs were the writers, translators, scribes, and all the earliest printing presses, as well as common twentieth century methods of mass communication. They controlled what could be said and whose speech could be heard. But, information technology and the gradual expansion of the Internet loosened their monopolized grip on the flow of information. In fact, the publication of this book is possible only because of these advancements.

The heart of our persecutors' domination machine is their fiendish, worldwide occult banking system, which is bound in place as much by faith in God as fear of damnation, since it would be impossible to enforce without coercion and threats. By promoting faith and trust in God on the part of the masses, they are able to commit their depraved crimes with a mandate from the faithful, who believe that allowing their corruption to flourish will hasten the coming of their ruler, Christ.

Without their "In God We Trust" financial system, they would be powerless to commit every other one of their innumerable crimes. It is by means of this core fraud that they control numerous subsystems, which return nourishment to the heart of their infernal machine. But, if those who trusted in authority lost their faith, the heart would fail, and the entire nefarious structure would crumble. Fortunately, information technology now presents a new opportunity to, at least, circumvent the heart of their domination machine and, if it is used properly, it has the potential to end their terrible reign.

This book is for witches, who are both well-suited to this new development and most in need of it, because many of us are concerned about how we can go about our business without interference. I wrote it because what is presented here is a complex combination of subjects and it is not enough for me to simply say, "Here is an opportunity to mitigate some vague harm that is being done to you," and expect that assertion to be accepted at face value. It is necessary to show you the exact nature of the crimes they have committed against us and how the remedy presented herein works to antidote the poison, which until recently we have all been forced to drink.

The history of their atrocities must be examined because, after centuries of colonization and witch hunts, their deceptive influence runs through every aspect of modern life. Their power is now so absolute, so ubiquitous, that it is often difficult to tell their manufactured illusions apart from the natural landscape. Therefore, we must analyze mundane things to determine their authenticity.

This books' purpose is to describe the architecture of their wholly unnatural constructs and their methods of domination, so you may fully recognize genuine avenues of escape and more easily perceive their bloody fingerprints on the traps they are setting for those who fail to make the distinction. Escape from the prison they have spent literally ages constructing is by no means a *fait accompli*; there are things that must be understood, which may require a mental adjustment. Those who fail to make the adjustment will continue to suffer.

To defeat an enemy, you must know him well and understand his psychology. When you are vastly outnumbered by enemies much more powerful than you are, then you must always work in the darkness, in secrecy. You must counter the occult with the occult. There is now a way to subvert your most powerful, dangerous, and ancient enemies while cloaked in the glorious blackness of the night. This little book is here to show you how to do it.

— Sophia diGregorio

INTRODUCTION

As witches, we reject authority and hold in the highest esteem the pursuit of occult knowledge. This puts us at odds with those who seek to dominate and control, who make an assault on knowledge and those who possess it. These self-ordained authorities withhold factual information while they propagate misleading information, and promote trust in authorities, thereby inducing mass delusion. The keystone to their present, unnatural world order, the one over-arching system that holds all of its insidious subsystems in place, is their financial system, which is a mystery to most people and, like God, taken on faith and rarely questioned.

The traditional banking system is a highly secretive occult order, an ancient religious brotherhood, referred to throughout this book as "the Holy Brotherhood," whose innermost circle of members have sought for millennia, to dominate and financially enslave the world. Money is central to their system of domination because it is a measure of your time, creativity, knowledge, skills, and life force energy. The banking system is the work of the worst black magicians on earth, religious supremacists, who believe they are ordained by God to rule the earth, to enslave unbelievers, and take whatever they want from us. By means of it, they suck the life essence from their victims by day and night.

The modern financial system is a mass surveillance operation, which the Holy Brotherhood uses to track and monitor you, your location, and your personal activities, including what books you read, your politics and religion, your friends, family, and associates. In fact, if you live in the United States, it is likely that one government bureau or another working in conjunction with numerous corporations have compiled a dossier on you.[1] The financial system is a pivotal component of the surveillance

1

state, which now exists in most industrialized nations. By means of this system, they cripple those who, in their estimation, have wrong thoughts and practices by denying them full participation in the economy.

From a witch's perspective, the world has changed little apart from increased subtlety on the part of the Inquisitors. In modern nations, these authoritarians are no longer dragging us from our homes and placing us in stockades, nonetheless, the goodwill of God's faithful is not enough to ensure either equal protection under the law or fair economic opportunity for witches. The only way to guarantee that your inherent rights remain untrampled is to exercise your basic right to privacy. By doing so, you entirely remove their ability to commit crimes against you. All people need privacy and historically disliked people need even more of it, financial privacy being its most fundamental form.

This book explains why the traditional banking system is especially problematic for witches, and why witches and occultists may be especially well-suited to using a heretical system of transaction. It describes the philosophy of Bitcoin and how Bitcoin and altcoins can best be used to increase your personal and financial privacy.

Bitcoin is not just another way to send or receive a payment, nor is it a speculative digital commodity. It is an important tool for subverting the main power structure of the harassing, omnipresent Holy Brotherhood, which they and their servants have so long used to dominate us. When properly used, it provides immediate relief from the strangling grip of your historical enemies, with the long-term potential to completely undo them.

This book begins with a history of the Holy Brotherhood's financial crimes, which are central to their centuries of other atrocities. It provides the reader with an unorthodox view of banking history, an analysis of the architecture of our historical enemy's system of domination, all the information you need to clearly discern the escape route from the saturnine, virtual prison cell into which we are all born, and, finally, instructions for getting yourself free of it all—*right now.*

CHAPTER 1
THE GREATEST SWINDLE EVER SOLD

From antiquity, money and religion have been intertwined. Imagery and statements displayed on coinage and currency demonstrate the relationship between money and local gods, the local rulers as gods or their agents, and religious ideology. Now, as in the past, modern bankers promote religious faith and trust in the authority of God and his elect.

Beginning, at least, four thousand years ago during the reign of King Hammurabi in Babylon, the banks were operated by religious men who charged fees to governments to mint coins, collected tithes, and operated currency exchanges from their temples. They kept people's real money, usually gold, in their vaults, while issuing IOUs to their account holders. Eventually, they began lending out the gold in their vaults and charging interest, at least, to their enemies. Then, they began lending more gold than they actually had in their vaults. This practice soon led to these swindlers issuing worthless IOUs for non-existent assets, which are now traded as currency.

The Central Banking Network

The pivotal component in their fraud, that which has provided them the ability to dominate distant nations on a worldwide scale for the past few centuries, is the concept of the central bank, which is no ordinary kind of bank. A central bank differs from a common bank in that it is a monopolized, private, foreign-owned corporation positioned like a parasite inside a country's government. It masquerades as an organic part of the legitimate government and is permitted to make usurious loans to that government, for which the citizens of the infected country are taxed to pay the interest, thus creating the "national debt."

The concept of citizen-owed debt exists under a different name everywhere a central bank exists and wherever there is one, there is always poverty, ever-increasing governmental and personal debt, inflation, and a state of ongoing economic crisis, which are all caused and managed by the central bank, itself. For example, in Western Europe, this circumstance is called the "European debt crisis," the "Eurozone crisis," or the "European sovereign debt crisis," and the alleged debt is owed to a foreign entity, the European Central Bank. Each central bank is a component in the worldwide network of central banks, which comprises the monolithic, privately owned, international banking system.

Article 1, Section 8 of the U.S. Constitution charges Congress with the responsibility "To coin Money, regulate the Value thereof, and of foreign Coin, and fix the Standard of Weights and Measures" at the Federal level. Instead of carrying out this duty, they abdicated it to a private, foreign-owned central bank, alternatively called the "Federal Reserve Bank" or the "Federal Reserve System," and nicknamed "The Fed." This monopolized international corporation is combined with a number of government bureaus, including the Federal Treasury and its military arm, the Internal Revenue Service. It includes a board, which independently decides the interest rate the citizens must pay.

The U.S. Federal government borrows money from the private Federal Reserve Bank at a usurious rate. The income taxes American citizens pay to the Federal Treasury only go to pay *the interest* on the government's loan from "the Fed," which is in turn paid to its international shareholders. These taxes do not, as many Americans believe, pay for infrastructure and government salaries. Instead, every April 15th, Americans pay interest on the government's "national debt" to a privately and internationally-owned bank. Every second of every minute, day and night, this fraudulent debt grows larger. Moreover, it is designed to increase infinitely, and so that your labor and property are progressively devalued through inflation, thereby siphoning off your personal wealth and the wealth of the entire nation to enrich the owners of the central bank.

Central Banking and the Lending of Fictional Treasury Reserves

In the United States, it is permitted for any bank, including the central bank, to lend ten times more money than it has in its vaults, which is a legally sanctioned, fraudulent practice known as "fractional reserve lending." It is unknown what the vaults of the Federal Reserve Bank contain because attempts by elected representatives to obtain this information by means of an audit have failed, and the reports they have presented to the public for decades "lack credibility."[1] There are no external oversights on the practice of fractional reserve lending. You are simply expected to take it on faith that the stated amount of gold reserves exists. Although if the Holy Brotherhood's history serves as an example, it is likely that their great treasury of reserves is comprised of nothing more than thin air.

Because the swindlers lend fictional reserves at usurious rates, the debt a government owes the central bank can never be paid off. It is mathematically impossible to do so because, in this scheme, at least ten times *more debt than currency to pay it* is constantly created.

To illustrate, if the vault contains a million dollars and the bank lends out ten million for mortgages, construction loans, business loans, car loans, and so on, this will lead to inevitable failure on the part of the majority of individual borrowers because the dollars necessary for 90% of them to pay back their loans do not exist. Only one in ten of those borrowers will be able to pay off his or her loan; the other nine are mathematically destined to default on their loans and the bankers will, thereby, obtain the assets named in the agreement. Of course, the borrowers ability to make the payments to the bank are further impeded by the burden of being charged interest.

It is important to understand that upon a borrower's default, bankers do not "take back" an asset, as is commonly said, but instead *acquire it*. Once a central bank is installed, this legalized swindle of lending fictional money to obtain real assets is the Holy Brotherhoods' primary method of dispossession, of acquiring the property of individuals and the entire host country by fraud.

God's Counterfeiters

A glance at the dollars you have in your possession will reveal that they are not actual money, only debt notes. Each one represents an ever-multiplying portion of the usurious national debt. When you exchange dollar bills for goods and services, you are circulating debt and an empty promise by a central bank, backed by faith and trust in a hostile, fictional God, and administered by the Holy Brotherhood.

The central banking system is a swindler's scheme, which is rigged to fail and, when it does, to bankrupt an entire nation, enslave individuals, and rob them of all their possessions by the creation of fraudulent national and individual debt.

Financial Abuse in the Name of God

In the macrocosm as in the microcosm, once an abuser has control over your finances, he can dominate, enslave, even destroy you and, in fact, over the centuries entire nations have been enslaved and destroyed by the fraud of the Holy Brotherhood. Their central banking system is a Black Lodge,[2] a secret society, a brotherhood of black magicians, an insidious Abrahamic cult of murderous thieves who harbor a bitter hatred of women and heathens, whom they believe are animals existing like livestock only to serve them. They say their supreme God has commanded them to exercise dominion over the world, to dispossess and enslave its people—specifically pagans, and because they imagine themselves as superior beings, they are able to justify acts of remarkable cruelty against us.

As it turns out, the Holy Bible is not just a chronology of thousands of years of pornographic violence and depravity by the Holy Brotherhood at the behest of their imaginary master. It is, also, a how-to guide for financial abusers. Their holy books are crime manuals, curiously preoccupied with matters concerning accounting, taxation, real estate, lawyering, and finance.

The Age of Pisces: The Holy Brotherhood's Two Thousand Years of Colonization

Christianity is a paganized descendant religion of Judaism, based on the life of the character, Jesus, a political insurrectionist whose birth roughly coincides with the beginning of the Age of Pisces. In the fourth century, Rome adopted Christianity as the Holy Roman Catholic ("Universal") Church. In effect, the establishment of this new form of Judaism in Rome was revenge for the Roman invasion of Jerusalem. This violent, religious form of government, based on the doctrines of the earlier Jews and Jesus, who is preposterously referred to as the "Prince of Peace," is the vehicle the Holy Brotherhood used to colonize Europe and later the Americas and other continents.

For nearly two thousand years, they have terrorized and dispossessed the natives, in particular women. It is their custom to destroy the native spiritualism, erect their own religious structures on top of their victims' burnt bodies, then demand the survivors worship their vengeful God, all while claiming that it is *they* who are being persecuted. They replace the native society with their own perverse social order. They plunder their victims' natural resources, whether silver, gold, oil, wildlife, or human labor and creativity, while they dominate them through fraud, violence, threats, and intimidation. They do this on the basis that their God commands them to do these things to foreigners and heretics, as written in their holy books.

Since overt theft and slavery are frowned upon, at least officially, they long ago devised and continue to perfect occult methods of robbing and enslaving their enemies, primarily through their banking system. Their aim is to concentrate all worldly wealth and power in their own hands and to enslave everyone who is not a member of their brotherhood.

The Holy Brotherhood's international network of banks, their worldwide central banking system, is the cause of most of the world's poverty. By means of it, they place limitations on both commerce and human potential, and restrict individual and national wealth. It is they who decide who can and cannot participate in their pernicious system and, as it stands, being denied participation is even worse than being forced to participate in it. The Holy Brotherhood and their banking system comprise the force behind most, if not all, wars and domestic conflicts. They are the cause of worldwide misery and the overall ruin of humanity.

Terms to Consider

The essential meaning of words, their origins, and present usage are especially important to this discourse, therefore, please, consider the following terms:

Dominate: From the Latin, "dominatus," which is the past participle of "dominari." "Dominari" means to rule, dominate, or govern, and is derived from "dominus," which means "lord," and "master." These words are related to the domination of the land, hearth, and home or "domicile." "Domus" means "home." To dominate is to lord over, to be the master of, to rule over, govern, or control.

El: An ancient word meaning "God," especially used to refer to the Canaanite god, "Yahweh." "Al," as in "Allah," is a variation on "El." "El" is the last syllable of the word, "Israel," which represents a trinity of powers, Is-Ra-El: Isis (Egyptian moon goddess and queen); Ra (Egyptian sun god and king); and El (the Canaanite El). "Ra," is present in the name "Abraham." In Hebrew, "Israel" means "Prince of God."

Regarding El, Madame Blavatsky wrote the following in *Isis Unveiled, Vol. 2.,* originally published in 1884:

> *El, the sun-god of the Syrians, the Egyptians, and the Semites, is declared by Pleyte to be no other than Set or Seth, and El is the primeval Saturn—Israel. Siva is an Aethiopian god, the same as the Chaldean Baal–Bel; thus he is also Saturn. Saturn, El, Seth and Kiyun, or the biblical Chiun of Amos, are all one and the same deity, and may be all regarded in their worst aspect as Typhon [Set] the Destroyer.* [3]

Related terms include "elite" (the followers of El, the *EL-ites*); "elevate" (to be raised up to God); "to elect" (to choose a representative of God); and "the elect" (those chosen by God, representative of God, God's Chosen).

Judge: Derived from the Latin noun, "judex," which means "judge," and the Latin and English word, "censor," meaning to "judge" or "moralize." The Latin verbs, "iudico" and "judico" mean to sentence, condemn, decide, adjudge, and adjudicate. The terms, "judge, "Jude" (meaning "Jew," a Hellenized form of "Judah," which is the name of an Israeli tribe), and "Judaism," are words related to Judeo-Christianity's brutal condemnation and judgment of others. God is the ultimate judge. In government, earthly judges possess the authority of God and do his work by passing secular judgment.

Heretic: From the Greek, "hairetikos," meaning "able to choose." It means one who is at variance with authoritarian standards.

Linguistic connections between the terms, "wealth" and "God," include the Latin words, "Divus," which is Latin for God, and "divitiae," which means "wealth," "affluence," and "estate." To the Holy Brotherhood, "godliness" means wealthiness. At their core, their only moral code is their endless quest for earthly riches, and the dispossession, domination, and murder of infidels in so doing.

Where the Holy Brotherhood are concerned, words and language are often intentionally manipulated to deceive outsiders. Inversion is their hallmark and an indication of their fingerprints on a matter. It is typical for them to invert the meaning of words, which, when they are reversed and set aright, reveal astonishing deceptions. For instance, it is common for them to say that they are being persecuted or robbed when it is, in fact, *they* who are doing the persecution and robbery; to say poisons are "medicine," that violence is "peace," and to call the forced participation in one of their fraudulent schemes a "choice." You will find many more examples of their inversions throughout this book.

CHAPTER 2
A CHRONOLOGY OF BANKING HISTORY WITH COMMENTARY

In creating this linear account of relevant events, the author has endeavored to obtain the most accurate information. Not all historical authorities may be regarded as reliable, especially when they are at variance with each other, and some appear biased or possess a clear motive for deception. Furthermore, the dates of plagues, wars, and historical ages or periods of time, are argued by religious scholars and secular historians. Nonetheless, this chronology with commentary demonstrates the interrelation of banking, corporations, religion, witch hunts, and wars and provides a context for the discourse of the chapters that follow.

c. 2300 B.C to c. 2000 B.C. The first banks known to exist were among ancient merchants in Babylon. Archaeologists believe that the first currency consisted of ingots and unmeasured metals. In Sumer, one of the earliest civilizations acknowledged by orthodox historians, archaeologists discovered "thousands of clay tablets inscribed with all sorts of Sumerian legal documents—contracts, deeds, wills, promissory notes, receipts, and court decisions."[1]

From antiquity, currency was minted, held at pagan temples, and lent to merchants by their priests. The concepts of taxation and indebtedness were used by both secular and religious leaders to confiscate land and other valuable property, and to bind individuals into slavery. Courts existed and were abused for this purpose, just as they are today. There was bureaucracy and corruption, at the root of which were priests, politicians, bankers, and their accountants. Since the officially recognized history of the world is narrow and incomplete, it is likely that banking and the exchange of currency in some form, has been a worldwide practice throughout many previous ages, the history of which has been hidden or lost.

1800-1400 B.C. It is claimed that the Jewish religion was formed at different time periods between these approximate dates. Some religious scholars say it was founded by Abraham, some say it was by Moses. There is no contemporary historical basis for either of these assertions outside their own religious books, therefore, Abraham and Moses may be mythological characters, much like Adam and Eve.

There is, also, interesting disagreement about where the Jews originated. It is generally agreed upon by historians writing much later that they came, at some point, from Africa, although some have asserted that they originated on the island of Crete. According to the famous, pagan historian Tacitus, the strongest belief at the time of his writing in 70 A.D. was that they had been expelled from Egypt into the desert, where they spent six days wandering in despair and would have died, if not for one among them, named "Moyses," who by trickery got himself designated as their leader, found water for them, and then, upon the seventh day, led them to another land, whereupon they expelled the natives and took their land for themselves.[2] But this tale appears mythological, too, especially in its preoccupation with the number, "7," which bears numerous associations to the planet Saturn, as further described by Tacitus.[3]

Furthermore, according to Tacitus, it was Moyses who, "wishing to secure for the future his authority over the nation, gave them a novel form of worship, opposed to all that is practiced by other men."[4] Everything loved by those around them became the subject of Jewish hatred and derision. Things regarded as immoral by others were regarded as sacred by the Jews, and vice versa. The scorn brought upon them by their "perverse and disgusting" customs solidified their nation.[5] He says, "...among themselves they are inflexibly honest and ever ready to shew compassion, though they regard the rest of mankind with all the hatred of enemies."[6]

The first God of the Jews was "El," from the Hebrew "Elohim." El has been syncretized with a number of other pagan gods associated with storms and fire, including Yahweh, the fire spirit to whom the Temple of Solomon was dedicated.

957 B.C. According to the *Old Testament,* the First Temple, also known as the "Temple of Solomon," was built on the notorious Temple Mount in Jerusalem. It was said to house the legendary Ark of the

Covenant, a big box, which supposedly contained the stone tablets upon which God engraved the Ten Commandments. Neither the tablets nor the Ark has ever been found. According to some accounts, the temple was destroyed by Nebuchadnezzar II after the Siege of Jerusalem of 587 B.C., although some Jewish sources claim it stood a few hundred years longer.

Allegedly, they soon rebuilt the temple on the same spot, calling it alternatively "The Second Temple" or "Herod's Temple." According to Tacitus, when the Romans, led by Titus, entered to destroy the Second Temple, which was said to house a great treasury likely comprised of tithes and taxes, they found it empty.[7] This and subsequent Roman incursions devastated the Second Temple and, by most accounts, it was completely destroyed. In c. 130 A.D., the Roman Emperor Hadrian erected a temple dedicated to Jupiter in its place.[8] The Jews were dispersed after the destruction of the second temple and did not return until the mid-20th century.

It is difficult to obtain archaeological information about either of these Jewish temples or even to determine their exact locations, due to erosion and changes in the terrain, and since Jewish religious leaders have designated the supposed locations "sacred," as a pretext for limiting any research. To further complicate matters, a lot of deadly violence occurs at the Temple Mount. Although there is historical evidence of both a First and Second Temple having actually been constructed somewhere in Jerusalem.

The Temple Mount remains at the center of worldwide religious and state discord. Presently, the faithful members of the Sinister Trinity, believing they are in their "End Times," await the construction of a third temple at this location, which is to be proceeded by the waging of a fiery world war and the subsequent coming of their invisible overlord to dominate the entire planet.

c. 700 B.C. to 601 B.C. The earliest recognizable coins to be discovered by archaeologists were found in Turkey, in what was once Lydia, and were manufactured from electrum, which is a naturally occurring alloy of gold and silver.[9] The coins were measured into units of equal weight and minted using the punch method.[10]

344 B.C. The Temple of Juno Moneta, in which Roman coins were first minted, was constructed on the Capitoline Hill in the city center of Rome. Roman mints were associated with pagan temples, which, also, housed legal manuscripts. This is an example of the relationship of pagan religion, finances, and government in what would become the power center of the Judeo-Christian religious state. The benefit to abusive rulers of associating government authority with religious belief, including belief in local spirits worshiped by pagans, was described in the sixteenth-century by Niccolò Machiavelli in *The Prince* and *The Discourses on Livy*.

323 B.C. to 31 B.C. During the Hellenistic period in Greece, banking often took place in Jewish temples, which allegedly housed their own treasuries. Only Jewish currency was accepted as a tax or tithe. Money changers were bankers who worked in or near the temples to exchange other currencies for Jewish currency for a fee. They, also, lent money for large business ventures, such as maritime endeavors, at very high interest rates of "22.5 or even 30%."[11] The first banks in Egypt, including the Royal Bank of Alexandria, were established by Greeks at about the same time.

c. 200 B.C. The Code of Manu (or "Laws of Manus," also, called "The Remembered Tradition of Manu" and "Manava-dharma-shastra," which means "The Dharma Text of Manu") records in Sanskrit the earliest known laws regarding mortgages.

380. Catholicism became the state religion of the Roman Empire. "Catholic" means "universal," a term which belies their intent for world domination. Even in the early days, they were far from being peaceful or respectable. Early Roman Christians are described by Tacitus, as criminals and perverts, and by the influential nineteenth-century occult scholar Heckethorn, as contentious "rabble," who "when not flying at one another's throats, were ever busy in spewing forth their fanatical venom upon all not of their ilk."[12]

609-632. According to Islamic scholars, the religion began when their founder, Muhammad, wrote the first words of their main doctrine, the *Koran*, alternatively spelled, "*Quran*." It is a synthesis of Judaism and local pagan worship, which is evident in its symbolism. Its crescent moon is the symbol of Hubal, the moon god whose statue was used for divination at the Ka'aba, in Mecca before it was destroyed along with other pagan idols by Muhammad in 630. Like Jewish and Christian history, Islamic history appears to be a combination of Judaic and local pagan mythology, with little evidence to support the alleged events outside their own doctrines. The word, "Islam," means "submission to God."

685 to 691. The Dome of the Rock was built at the Temple Mount site, by the caliph Abd al-Malik ibn Marwan, as a shrine. It is the location from which they claim their founder, Muhammad, ascended to Heaven. Its exact location is believed to be the site of the Temple of Solomon and the Temple of Jupiter Capitolinus. They forbid its entry to religious Jews, relegating them to praying along an outer wall, called "the Western Wall," which the Jews believe is a portion of the Second Temple, despite records stating that the structure, wherever it once stood, was long-ago obliterated.

1288 to the Present. Stora Kopparberg Bergslags Aktiebolag in Falun, Sweden, which believes itself to be the oldest, still-existing joint-stock company, is known to have had an active copper mine around the year 1000. A document from 1288 shows that the Bishop of Västeras owned a share in the mine.[13]

1095 to 1102. Pope Urban II initiated the First Crusade to keep access to the Holy Land and fend off the Muslim Turks who sought to control it. Europeans captured Jerusalem in 1099. Afterward, the Holy Brotherhood's Sinister Trinity of religions fought among themselves for control of their Holy Land. In Europe, the Crusades were funded through both tithes and taxes, as well as the confiscation of property from heretics.

c. 1118 to 1312. The Knights Templar, also known as "The Poor Fellow-Soldiers of Christ and of the Temple of Solomon," the "Order of Solomon's Temple," and simply "The Templars," existed as a Catholic military order, an army of God. More than just a holy order of the Judeo-Christian Brotherhood, the Templars were ruthless plunderers and bankers who provided a range of financial services, including lending, pawn brokering, and issuing traveler's checks to their customers. On Friday, October 13, 1307, King Philip IV of France had them arrested and tortured, seized their property, and had them burned at the stake on the grounds of heresy, apparently in order to avoid repaying money he had borrowed from them. The pope officially dissolved the Order in 1312.[14]

1147 to 1149. The Second Crusade. Announced by the pope and led by many kings of France and the German lands, it was mostly a failure on the part of the Christians.

1184 to c. 1230. The Medieval Inquisition. A series of papal and episcopal inquisitions aiming to snuff out heretics.

1187. Muslims regained control of Jerusalem. By 1291, they completely controlled the Holy Land.

1188 to 1192. The Third Crusade. Christians tried and failed to take control of the Holy Land.

1251 to 1320. The Shepherds' Crusades. Again, the Christians tried and failed.

1254. Pope Alexander IV established the Office of the Inquisition in Italy.

1300 to 1600. The Renaissance. Italy established many banks, especially in Venice, Genoa, and Florence during this time.

c. 1347 to c. 1359. The Black Death, a plague killed about one-third of the population in Western Europe. This and subsequent plagues were blamed on witches conspiring against Christians. The Holy Brotherhood's war on witches gathered more steam.

1372. *Directorium Inquisitorum* was written for the Inquisition by Nicolau Eymeric, a Dominican in Spain. It is the basis for many later witch hunting manuals.

1401. The *De Heretico Comburendo*, a statute passed by King Henry IV of England punished heretics by public burning at the stake.[15] Its passage stemmed from concerns about "increasing criticism of clerical wealth."[16] The law specifically censored heretical books and writing.

c. 1439. The Gutenberg printing press was introduced. Invented by Johannes Gutenberg, its first publication was a version of the Holy Bible. Soon after, it was used to produce books on demonology and manuals for witch hunters.

1458. The first known mention of double-entry accounting was made in *The Book on the Art of Trading,* by Benedikt Kotruljevićin. The techniques of accountants, tax collectors, and money lenders, became increasingly sophisticated during the years of the crusades. Modern fractional reserve lending would be impossible without the development of double-entry accounting.

1472. The oldest living bank in the world, Banca Monte dei Paschi di Siena, headquartered in Siena, Italy, was established.

1478. King Ferdinand II of Aragon and Queen Isabella I of Castile established the Tribunal of the Holy Office of the Inquisition, more simply known as "the Spanish Inquisition." Heckethorn wrote, "From the earliest days of Christianity the Inquisition existed in the spirit, if not in the form,"[17] however, this version of it differed from any in the past in that it was the first secular witch hunt.

1487. The *Malleus Maleficarum,* the most notorious instruction manual of the Inquisition and a cruel libel against witches and women, was published in Speyer, Germany, by Heinrich Kramer and Jacob Sprenger, Catholic clergymen of the Dominican Order. This guide to our torture and extermination was used by secular courts throughout Europe. It was very popular and sanctioned by the pope. While most Christians and secularists now publicly disavow such words and deeds, nonetheless, the sentiments of this book and its authors live on.

1492. Columbus set out on his voyage from Europe to discover what is now called The Bahamas. Officially, the purpose of this trip was to establish more expeditious international trade routes; less officially, it was to expand the influence of the Spanish rulers, to con, conquer, and enslave, and to teach more pagans to trust in God at sword point. This legendary voyage preceded the Holy Brotherhood's centuries of bloodshed, rape, robbery, fraud, enslavement, and the death and destruction of entire nations of the earliest Americans in the name of God.

Soon after his arrival, the Spanish Inquisition, established by Columbus' generous benefactors, Ferdinand and Isabella, was dispatched to Mexico where it was sometimes known as the Mexican Inquisition. Catholics, in particular the Dominicans and Jesuits, also, carried out atrocities in Spanish and French territories that would later become part of the United States, as well as in Central America and parts of South America. Their pretext for robbing the New World of gold, silver, and other resources was that our land was owned by their God and his earthly representatives, therefore, the natives owed God a tribute.

1532. *Constitutio Criminalis Carolina.* The Holy Roman Emperor Charles V decreed that harmful witchcraft was punishable by death by fire. Witches who caused no harm were not burned, only imprisoned.[18]

1542. *The Witchcraft Act.* The notorious wife-killer, King Henry VIII, made witchcraft a crime against the state by enacting the first Witchcraft Act. Under this law, convicted witches were put to death, and the state confiscated their property. There were numerous similar laws in Scotland, Wales, and Ireland, and a long succession of additional Witchcraft Acts proceeded this one.

c. 1560 to c. 1580. The southern German lands were besieged by witch hunters. The accused and condemned were primarily women and, in a few instances, they tortured and murdered every woman and girl in entire villages. Some historians theorize that this particular episode of witch hunting fever was brought on by a few years of unfortunate weather for which they blamed witches.

1572. Augustus, Elector of Saxony, decreed that all witches should be burned, even mere diviners.

c. 1580 to c 1650. Most historians believe this is the period in which the greatest number of accused witches were murdered in Western Europe.

1597. *Demonologie,* the infamous treatise on witchcraft written by King James I of England, also known as King James VI of Scotland, was published. James had become intimately involved in Scottish witch persecutions.

December 31, 1600. The British East India Company was granted a charter by England's Elizabeth I. This company had its own large army, which it deployed in India. It ceased to exist on June 1, 1874.

1602. *Discourse on Witches* was written by Henri Bouget, who conducted numerous witch trials and condemned many to death in France where this witch hunting manual was commonly used.

March 20, 1602. The United East India Company (also known as "The Dutch East India Company"), headquartered in Amsterdam, was chartered by the Dutch government and traded in tea and spices. It is widely considered the first international corporation and the first in the world to publicly issue stocks and bonds. It went out of business on December 31, 1799.

1604. *The English Witchcraft Act of 1604, An Act Against Conjuration, Witchcraft and Dealing with Evil and Wicked Spirits,* was enacted by King James I of England, after which English witch persecutions intensified.

1644 to 1647. Witch-Finder General Matthew Hopkins enforced the *Witchcraft Act of 1604,* murdering about 400 accused witches mostly in East Anglia. Upon his retirement in 1647, his book, *The Discovery of Witches—In Answer to Several Queries, Lately Delivered to the Judges of Assize for the County of Norfolk,* was published in London.

1668. The first central bank, Riksbank (in English, "National Bank"), was established in Sweden.

February 1692 to May 1693. The Salem Witch Trials, a series of persecutions, hangings, and murders of accused witches, took place in the British colony of Massachusetts.

October 1692. Harvard College alumnus, promoter of allopathic medicine, scholar, scientist, and Puritan minister Cotton Mather published his book, *The Wonders of the Invisible World,* in which he condemned witches and witchcraft. The book, first published in Boston, proved popular and, at least, two reprints quickly followed its first release, including one by a London press in 1693.

July 27, 1694. The Bank of England, a central bank, was established by Royal Charter to finance a war with France. Formed as a private corporation with private shareholders, it is the model for all modern central banks. *The Magna Carta* previously placed limits on usury, however, this protection was lifted in the same year.

1735. *The Witchcraft Act of 1735* (9 Geo. II c. 5) was passed by the Parliament of the Kingdom of Great Britain making it a crime for a person to claim that any human being possessed magical powers or was guilty of witchcraft. Thus, the witch hunts were officially abolished in Great Britain, however, witches were still imprisoned for "pretending" to practice witchcraft.

1750 to 1764. Colonial Scrip, interest-free paper currency, fiat and not backed by gold or silver, was issued by the American Colonies.

1764. England, under King George III, outlawed all forms of paper money issued by the American Colonies and forced them to use usury-based currency issued by the Bank of England. Once thriving and wealthy, the people in the colonies soon became impoverished.

1765. *The Stamp Act of 1765* (*Duties in American Colonies Act 1765*; 5 George III, c. 12) imposed a direct tax on the Colonists. Some historians regard this as the beginning of the American Revolution.

July 4, 1776. The United States of America was formed out of the original thirteen British colonies after a war instigated by George III and the centralized Bank of England. What followed has been a continual struggle by the Holy Brotherhood to dominate and enslave the people of the United States and other regions of the world by means of their domination machine.

May 26, 1781. The President, Directors, and Company, of the Bank of North America, commonly known as the "Bank of North America," a private bank, was chartered by the Confederation Congress as the first de facto central bank in the United States. It was re-chartered in 1787, preventing it from fulfilling its role as a central bank. It later merged with Wells Fargo. The Bank of North America was the first corporation in the United States. to make an initial public offering (IPO) of stocks.[19]

September 3, 1783. The signing of the *Treaty of Paris* formally ended the American Revolution.

1790. Philadelphia, Pennsylvania served as the temporary capital city of the United States of America. while Washington, D.C. was being constructed. The previous capital had been New York City.

July 16, 1790. Washington, District of Columbia became the new capital city of the United States of America. It was formed out of portions of the states of Virginia and Maryland, which were previously occupied by the Algonquin-speaking Nacotchtank. Capitol Hill stands on the spot where the old town of Rome, Maryland once was. Washington, D.C. is nicknamed "Rome on the Potomac." It was redistricted in 1871 by means of the Organic Act of Congress.

February, 25, 1791 to 1811. The First Bank of the United States was founded in Philadelphia, Pennsylvania. Chartered for a term of twenty years, it succeeded the Bank of North America in its role as the de facto central bank of the United States.

June 18, 1812 to March 23, 1815. The War of 1812. This was another war against the British, which occurred when they tried to block trade. It is sometimes regarded as a second war of independence. By the end of it, the U.S. government was deeply in debt.

1816 to 1836. The Second Bank, yet another central bank, was chartered for 20 years under the leadership of President Madison.

April 12, 1861 to May 9, 1865. The U.S. Civil War. During the early years of Lincoln's presidency, the U.S. issued genuine fiat currency, called "greenbacks." In 1863, when Union forces desperately needed more money to win the war, the *National Bank Act* was passed and afterward the money supply was, again, issued by central bankers and the new currency was created out of usurious banking debt. In 1864, the words, "In God We Trust," first appeared on a two-cent coin issued by the privately-owned central bank.

July 1, 1862. Lincoln and Congress "created the position of Commissioner of Internal Revenue and enacted an income tax to pay war expenses. The income tax was repealed 10 years later. Congress revived the income tax in 1894, but the Supreme Court ruled it unconstitutional the following year."[20]

February 3, 1913. Despite the fact that all previous attempts to establish a Federal income tax had been denied on ethical and Constitutional grounds, the U.S. Constitution was amended with the ratification of the 16th Amendment, which gave "Congress the authority to enact an income tax. That same year, the first Form 1040 appeared after Congress levied a one percent tax on net personal incomes above $3,000 with a 6 percent surtax on incomes of more than $500,000."[21] Between 1918 and 1953, the income tax collection bureau of the Department of the Treasury was sometimes referred to as the Bureau of Internal Revenue. In 1953, it was formally named the Internal Revenue Service.

December 23, 1913. The Federal Reserve Act was illegally enacted without proper ratification and signed by Woodrow Wilson. This act established the Federal Reserve System, including the private central bank, "The Fed." This treason was the result of plans hatched in secrecy during meetings held by bankers, including "Frank Vanderlip, president of the National City Bank of New York, Henry P. Davison, senior partner of J.P. Morgan Company, and generally regarded as Morgan's personal emissary; and Charles D. Norton, president of the Morgan-dominated First National Bank of New York."[22]

The Federal Reserve Bank, "The Fed," is a private corporation headquartered at the Eccles Building in Washington, D.C. and comprised of twelve regional central banks. It offers shares, but only to other banks within the international network of central banks. According to Eustace Mullins in the Foreword of his book, *Secrets of the Federal Reserve,* he was able to trace the original stockholders back to London, which he terms the "London Connection." Congressman Louis McFadden described a similar connection in his 1934 statement to Congress.[23] In fact, every central bank that has ever existed in the United States has a British connection involving the reigning monarch at the time or a foreign-owned central bank in England.

July 28, 1914 to November 11, 1918. The United States entered World War I with Woodrow Wilson as president. Under the new Federal Reserve System, billions of U.S. citizens' hard-earned dollars were lent to the Allies and never repaid, however, great sums of interest went to bankers in New York.[24]

January 10, 1920 to April 20, 1946. The League of Nations was formed a result of the Paris Peace Conference. The stated aim of the organization was to promote world peace.

1921. The Council on Foreign Relations (CFR), a nonprofit think tank, headquartered in New York City, with an additional office in Washington, D.C., was founded to direct all aspects of the Holy Brotherhood's establishment following the end of World War I. Its members include politicians, directors of intelligence agencies, bankers, lawyers, top university staff, and elite members of the media.

October 29, 1929 to 1939. The Great Depression, an engineered run up then sudden drop (a "pump and dump" scheme) in the value of corporate stocks, during the presidencies of Franklin D. Roosevelt and Herbert Hoover, led to economic disaster, not just in the United States, but in a number of countries.

September 1, 1939. Hitler invaded Poland. Historians generally agree that this event began World War II in Europe.

December 7, 1941 to September 2, 1945. Japan bombed the United States Navy base at Pearl Harbor on the island of Oahu, Hawaii, providing Franklin D. Roosevelt with the impetus for bringing Americans into the fray of World War II.

October 24, 1945. After months of planning, the United Nations was established to replace the League of Nations and fulfill their ostensible mission of world peace. The U.N. charter was drafted and signed during the Bretton Woods Conference, more formally called the United Nations Monetary and Financial Conference, which was held in San Francisco. This meeting of high-ranking politicians and high-level financiers spawned the International Monetary Fund (IMF), the International Bank for Reconstruction and Development (IBRD), the International Development Association (IDA), and the World Bank. The United Nations is headquartered in New York, New York and the other four are headquartered in Washington, D.C. These organizations were to lend and oversee the lending of the ill-gotten gains of the central banking financiers who designed World War II, as they further debt-enslaved those nations in the name of reconstruction. All of the many wars the

United States engaged in after World War II and the establishment of the United Nations have been illegal under U.S. Constitutional law since war must be declared by Congress.

May 14, 1948. The State of Israel was established and U.S. President Harry S. Truman recognized it on the same day. The formation of this new nation grew out of the adoption of a United Nations General Assembly resolution of November 29, 1947. It was one of the first acts taken by the newly established United Nations.

1950 to 1953. Korean War.

1951. All *Witchcraft Acts* were repealed in England and replaced with the *Fraudulent Mediums Act of 1951*. The repeal led to a flourish of public, commercialized witchcraft in England, which was exported to the United States and around the world, breathing new life into interest in the occult. In 2008, the new law was, also, repealed and replaced by a consumer regulation, *The Consumer Protection from Unfair Trading Regulations 2008*.[25]

Although these Acts, which strictly forbade the practice of witchcraft and made even pretending to do so a serious crime, there are still laws that have a similar effect. Today, in the United States and throughout Britain's Commonwealth of Nations, anti-fraud, anti-fortunetelling, and anti-conjuring laws aimed at witches and occultists still exist. Zoning regulations, licensing laws, biased legal definitions of religion and medicine, and legal restrictions on the gathering, growing, and commerce of certain herbs are used to restrict the traditional practices of witches. Witchcraft has been decriminalized in name only.

July 30, 1956. A law passed by the 84th Congress (P.L. 84-140) and approved by Eisenhower declared the phrase, "In God we trust," to be the official national motto of the United States.

March 8, 1965 to April 30, 1975. The Vietnam War. The United States entered the fray of a war begun in 1955 by France in a far off Asian nation, which most Americans had never heard of before. The very unpopular draft was ended under Nixon on January 27, 1973, but the U.S. military continued to send Americans into combat there until 1975.

June 5 to June 10, 1967. The Six-day War was fought between the State of Israel and its neighbors, Jordan, Syria, and Egypt, which was then called the "United Arab Republic." As a result of this war, it was decided that Jerusalem would remain an international city, accessible to all members of the three Abrahamic religions. Despite this sentiment, Jerusalem remains the epicenter of ceaseless violence among factions of the Sinister Trinity.

1974 to 1975. Intervention in Angola.

1983. Invasion of Grenada.

August 20, 1985 to March 4, 1987. The Iran-Contra Affair was the scandal that developed from illegal activities beginning in 1970s with the overthrow of the Nicaraguan President Anastasio Somoza. According to Somoza's memoirs, shadowy entities within the U.S. government fomented a civil war in Nicaragua, then financed both sides of the engineered conflict. 26 It was learned during the Iran-Contra hearings that individuals personally profited from supplying arms, running drugs, and likely even more heinous international criminal activities. The scandal went to the top of the Reagan administration and included the President, military generals, intelligence agents, and contractors. Only one of the men involved, a clandestine operative named Clines, ever saw the inside of a prison cell, although it was for failure to report all of his income to the IRS.[27]

Late 1980s to 1990s. Variable interest rates or "adjustable interest rates" were introduced to the American mortgage market with devastating results to the economy. Adjustable rates typically feature low rates upon introduction of the loan, then raise to high levels the borrower cannot pay. The issuance of adjustable rate mortgages (ARMs) to high-risk borrowers, along with internal banking fraud and forgery, played a large role in the securities-based mortgage crisis, which climaxed in 2008.

August 1990 to February 1991. Persian Gulf War. This U.S. led intervention in Iraq was the first of many similar actions.

1992 to 1994. Somali Civil War.

September 19, 1994 to March 31, 1995. Operation Uphold Democracy. A military intervention authorized by the United Nations Security Council to remove the leadership of Haiti.

1992 to 1995. Bosnian War.

February 1998 to June 11, 1999. Kosovo War.

September 11, 2001. A second attack was perpetrated on the World Trade Center in Manhattan a colossal complex of seven buildings designed as a place of world commerce and located in New York's Wall Street District, by actors whose identities are yet to be confirmed by any unbiased sources. A previous, less successful attack, occurred on February 23, 2003. This second event was used as the pretext to enter Iraq in 2003. Wars and interventions continue under the pretext of fighting Middle Eastern terrorism.

2001 to the Present. Afghanistan War.

2003 to 2011. Iraq War.

September 19, 2008. The Banker Bailout. When their mountain of banking fraud, involving mortgages, credit cards, and worthless securities, could no longer be concealed, President G.W. Bush proposed a plan to "bail out" the banks, called the *Emergency Economic Stabilization Act of 2008*. On October 1, the U.S. Senate passed an amended version of the bill, which cost taxpayers $700 billion through the *Troubled Assets Relief Program (TARP)*.

The bill was passed amid threats of military violence in the United States. [28] On October 2, 2008, Congressman Brad Sherman said, "Many of us were told in private conversations, that if we didn't pass this bill on Monday, the sky would fall, the market would drop two or three thousand points, another couple thousand the second day, and a few members were even told that there would be martial law in America if we voted 'No.'"[29] The banker bailout was a great transfer of wealth from individual Americans to the internationally-owned banks of the Federal Reserve System, resulting in long-term economic devastation and lingering uncertainty about the U.S. economy.

January 3, 2009. The greatest potential threat to the Holy Brotherhood, the true Antichrist, Bitcoin, was born when Satoshi Nakamoto mined the first block of its blockchain. The genesis blockchain code contains the following message: "The Times 03/Jan/2009 Chancellor on brink of second bailout for banks,"[30] an apparent reference to an article in a British newspaper, *The Times*, entitled "Chancellor Alistair Darling on Brink of Second Bailout for Banks: Billions May Be Needed as Lending Squeeze Tightens."[31] It is claimed that the founder of Bitcoin began writing the open source code in 2007, which was an especially dark year for the United States and other countries, due to a culmination of banking malfeasance. On August 18, 2008, one month before the passage of the treacherous banker bail out bill, the domain name, bitcoin.org, was registered.

February 15, 2011 to October 23, 2011. Libyan Civil War. A series of administrative shakeups occurred in their bank, The Central Bank of Libya, before, during, and after the conflict.

2014 to the Present. The atrocities of the Islamic State of Iraq and the Levant, also, known as Daesh, ISIS, and ISIL, have been the official reason for waging war in Iraq and a number of other Middle Eastern nations, and conducting mass surveillance in Western nations.

Conquest and Acquisition

Among other things, this chronology reveals the relationship of banking to war. At least, since the Crusades, the conquest and occupation of foreign nations is closely associated with banking and acquisition. Since the seventeenth century, it may be seen that a central bank is commonly established in a host country before, during, or after a war.

Modern wars and internal conflicts do not naturally occur because of racial or national hatred, nor conflicting ideologies, but are orchestrated behind the scenes by those with a commercial interest in them. They begin and end only at the pleasure of the Holy Brotherhood, who have spent 2,000 years concentrating their power and constructing their domination machine. Their fraudulent acquisitions are now so extensive, that they own the corporations that create the war machines, chemicals, and propaganda. They own the politicians and fund every side of wars and internal conflicts. Once a war is ended, they arrange to finance the reconstruction of their victims' nations at usurious interest rates.

Central banking, itself, is a form of foreign invasion and warring, accomplished by deceit. Every country, in which a central bank has been established, experiences economic depression, recession, ever-increasing inflation, poverty, homelessness, and ever-deeper indebtedness.

Since the establishment of the Federal Reserve System in 1913, the United States has been almost constantly involved in foreign wars and interventions under the pretext of fighting fascism, communism, or terrorism. In every conflict, the central bankers profit at the expense of all the nations involved and sometimes establish themselves within a nation where they did not previously exist. Presently, only three major countries remain without an internationally-owned central bank: Iran, Cuba, and North Korea.[32] Other errant countries have already been brought into compliance, with the exception of small governments, such as Vatican City, which operates its own private bank.

In every country, these international, private enterprises are named in such a way as to mislead the citizens to believe that they are part of their government, such as "National Bank" or "Federal Bank." But, their deceit does not end there. At least, in the United States, the official websites for information related to the history and nature of the central bank contain some blatant inaccuracies.

This is to be expected. Brazen mendacity is easy for them, since they regard themselves as ministers of the ultimate authority, and they have spent centuries taking control of nearly every aspect of people's lives, including education and media. After all, the official history is theirs, and it is virtually a model of deception.

The Holy Brotherhood's Domination Machine: The "In God We Trust" System

At the heart of the Holy Brotherhood's domination machine is their banking system. It symbiotically supports and is supported by numerous subsystems. The components of this diabolical machine are integrated and animated by blind faith, baseless trust, and authoritarianism, as expressed in their motto, "In God We Trust."

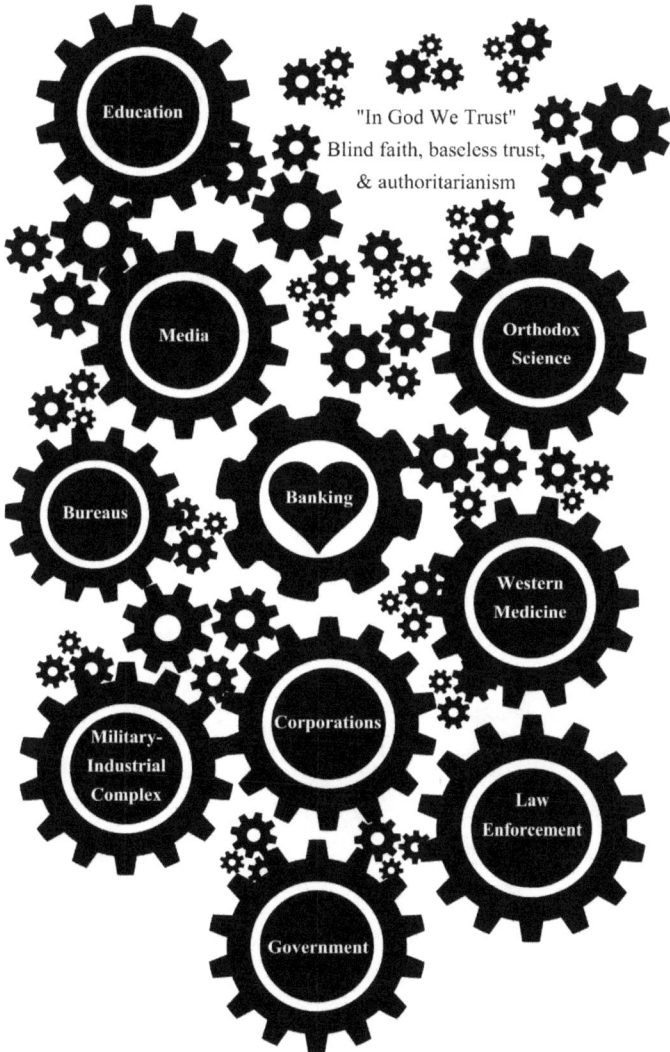

"In God We Trust"
Blind faith, baseless trust, & authoritarianism

Education

Media

Orthodox Science

Bureaus

Banking

Western Medicine

Military-Industrial Complex

Corporations

Law Enforcement

Government

CHAPTER 3
WHAT IS THE MEANING OF "IN GOD WE TRUST?"

"...the United States of America is not, in any sense, founded on the Christian religion..."
— Article 11, Treaty of Peace and Friendship between
the United States and the Bey and Subjects of Tripoli of Barbary, 1796.

In blatant mockery of the First Amendment of the Bill of Rights, which is supposed to protect citizens from the establishment of religion in government, "In God We Trust" was adopted in 1956 as the official motto of the first secular republic on the face of the earth, replacing the unofficial one, "*E pluribus unum.*" This same contemptible motto began appearing on all U.S. currency the next year after a mandate by President Eisenhower that it be placed on all currency and coinage in the United States.

Stamping this religious slogan on U.S. bank notes beneath the name of the country might violate the First Amendment except that the Federal Reserve Bank is a for-profit, private, corporate entity, which is owned by its international banking shareholders, rather than a government agency.

Much like a virus, The Fed pretends to be organic to the body of the Federal government, but the truth was revealed by Louis McFadden, Chairman of the House Banking and Currency Committee and Congressman in his statement to Congress in 1934:

> *Some people think that the Federal Reserve Banks are United States Government institutions. They are private monopolies which prey upon the people of these United States for the benefit of themselves and their foreign customers; foreign and domestic speculators and swindlers; and rich and predatory money lenders.*[1]

The Federal Reserve System was brought into being in the United States on December 23, 1913, under questionable circumstances. It is based on a combined government and religious system of usury and tithing, which is a religious tribute, from which the concept of taxation is derived. The Holy Brotherhood began developing this system thousands of years ago to deprive their enemies, as individuals and as entire nations, of their wealth and turn them into debt slaves.

Usury and Taxation

Usury is an ancient, dishonest commercial practice involving charging interest and fees to steal assets by means of loan agreements rigged against the borrower. If the borrower fails to make payments according terms of the contract, the lender takes the property named in the agreement.

"Usury" and "interest" are synonymous terms, however, rates of interest are differentiated as either illegal usury or legal interest under various state laws. Regardless, all interest, by whatever name, is usury in the common language. Biblical references to usury, also, refer to any interest.

Because charging interest on loans is a deceptive, innately criminal, and destructive practice, it is prohibited for members of the Holy Brotherhood to do it to each other, however, they are commanded by God to commit usury against their enemies, as described in *Deuteronomy 23:19-20:*

> *(American Standard Version) "**Thou shalt not lend upon interest to thy brother;** interest of money, interest of victuals, interest of anything that is lent upon interest. **Unto a foreigner thou mayest lend upon interest;** but unto thy brother thou shalt not lend upon interest, that Jehovah thy God may bless thee in all that thou puttest thy hand unto, **in the land whither thou goest in to possess it.**"*

> *(King James Version) "**Unto a stranger thou mayest lend upon usury;** but **unto thy brother thou shalt not** lend upon usury: that the Lord thy God may bless thee in all that thou settest thine hand to **in the land whither thou goest to possess it.**"*

The last phrase of this passage reveals the Holy Brotherhood's intentions when they charge usury, which is to possess the land of their enemies. Usury is a covert method of invasion by which they take other people's land and resources for themselves.

The collection of taxes to pay the interest on the national debt to the Federal Reserve Bank is usury. In accordance with their holy books, which forbid usury among the brethren, the Holy Brotherhood's churches, synagogues, and mosques in the United States claim "tax exempt status"
with the I.R.S. Likewise, they exempt large secular corporations owned and controlled by the brethren by means of tax "loopholes."

Coerced Credit

The Holy Brotherhood ordained a bishopric of three private corporations, Experian, TransUnion, and Equifax, whose office is to pass temporal judgment on your credit-worthiness. Seemingly omniscient, they collect personal and financial data on you, whether you have ever applied for credit or not. Based on this information, which is often false, they even rate you as if you are side of beef. Bankers may deny you a usurious mortgage, business loan, or credit card based on their bishops' reports on you.

Because of the inflation created by God's swindlers, credit is the only way most people can acquire the things they need. Moreover, through television marketing of credit card services, they have even managed to make usury appear prestigious. While these religious supremacists, who own the temples of finance, charge you substantial interest, they lend to each other at or near 0% interest because *thus saith the Lord their God.*

The Illusion of a Secular Republic: Romans 13

The fusion of God and state is a necessary component in the Holy Brotherhood's multi-faceted fraud. Because of it, monotheism, monopolies, hierarchies, authoritarianism, tithes, taxes, and usury are all abnormalities that long ago became part of daily life. Despite the prevalence of religiosity, God's existence is questioned far more often than his secular authority, whether governmental or corporate. As a result, even the most rigid materialist atheists tend to perceive the world through the cracked lens of God's authority.

The Holy Brotherhood easily maintain power by promoting submission to both God and the state as God. In the United States, which is the first secular republic in history, there are no kings or princes ordained by God to rule, but there are still executives, congressmen, senators, governors, judges, courts, and police, all of whom are a secularized version of God's ordained rulers. By means of these outwardly secular proxies of God, we still live under the religious law described in *Romans 13: 1-7,* which dictates the duties of citizens to secular authorities.

Two similar translations of *Romans 13: 1-2* are as follows with some phrases bolded for emphasis:

> *(King James Version) "Let every soul be subject unto the higher powers. For there is no power but of God: the powers that be are ordained of God. Whosoever therefore resisteth the power, resisteth the ordinance of God: and they that resist shall receive to themselves damnation."*

> *(The World Messianic Bible) "Let every soul be in subjection to the higher authorities, for there is no authority except from God, and those who exist are ordained by God. Therefore he who resists the authority withstands the ordinance of God; and those who withstand will receive to themselves judgment."*

In the above verses, the term, "higher powers," refers to government, "powers that be" are secular rulers; and "higher authorities" are politicians and monarchs. Resisting secular rulers is the same as resisting God and those who do so are to be damned or judged.

Romans 13:3-5 is especially loved by some law enforcement agencies across the United States because it urges you to place yourself under God's authority and that of his infallible minions, which may be interpreted as police, and threatens consequences in the afterlife for failure to do so. It reflects the disturbing, modern authoritarian view of police power and the individual's obligation to submit to law enforcement agents, regardless of the circumstances. *Romans 13* continues, as follows:

> *(King James Version) "For rulers are not a terror to good works, but to the evil. Wilt thou then not be afraid of the power? do that which is good, and thou shalt have praise of the same: For **he is the minister of God to thee for good.** But if thou do that which is evil, be afraid; for he beareth not the sword in vain: for **he is the minister of God, a revenger** to execute wrath upon him that doeth evil."*

> *(The World Messianic Bible)* *"For rulers are not a terror to the good work, but to the evil. Do you desire to have no fear of the authority? Do that which is good, and you will have praise from the authority,* **for he is a servant of God to you for good.** *But if you do that which is evil, be afraid, for he doesn't bear the sword in vain; for* **he is a servant of God, an avenger** *for wrath to him who does evil."*

The above, fork-tongued verse is an assurance that no good person needs privacy from authoritarians because if you are doing nothing wrong, there is nothing to fear from your violent overlords, since whatever they do to you is for your own benefit. The "ministers of God," referred to in the *King James* translation, are sometimes directly translated as "police," who are described as God's secular avengers ordained to execute his wrath. It may, also, be applied to judges, since theirs is the secularized judgment of God.

The terms "good work" and "good works" mean "good people." The use of the word, "good," in this context is, also, subjective since it relates to obedience to the Abrahamic God. "Good" people are Godly people. In other words, to do "good" is to aid and abet your historical enemies. "Good," as it is used here, does not refer to any common standard of ethical behavior.

Romans 13: 6-7 tries to guilt you into paying tributes and taxes to monetarily support your abusers, as follows:

> (King James Version) "Wherefore ye must needs be subject, not only for wrath, but also for conscience sake. For for this cause **pay ye tribute** also: for they are God's ministers, attending continually upon this very thing. **Render therefore to all their dues**: tribute to whom tribute is due; custom to whom custom; fear to whom fear; honour to whom honour."

> (The World Messianic Bible) "Therefore you need to be in subjection, not only because of the wrath, but also for conscience' sake. **For this reason you also pay taxes**, for they are servants of God's service, continually doing this very thing. Therefore give everyone what you owe: **if you owe taxes, pay taxes**; if customs, then customs; if respect, then respect; if honor, then honor."

The word, "customs," above, refers to taxes, tributes, and services owed to a lord by a peasant in Medieval Europe.

Romans 13 is one of our historical enemies' most egregious pronouncements, which demands total submission to unknown men and payment to them of monetary entitlements. According to this Judeo-Christian doctrine, secular government is ordained by God for his purposes, which include the domination and dispossession of heretics.

The Faith Based Initiative, an unconstitutional Federal taxpayer-funded program that encourages religion and obedience to God, promotes *Romans 13* in communities throughout the United States.[2] Some newly copyrighted translations of this passage, such as the *New Living Translation* and *The Living Bible,* even more explicitly interpret the ministering swordsmen as police who are infallible emissaries of God and government agents as God's equally infallible servants who are owed blind obedience, trust, and, of course, your money.

God, Gold and Government

The Holy Brotherhood dominates every aspect of public life. There is no known witch, pagan, or atheist of any kind in Congress, the U.S. Senate, the Supreme Court, or ever in the Executive Office of the President. Most U.S. politicians brazenly campaign on a belief in the Abrahamic God, hypocritically demonstrating such belief by offering public prayers to him. It is nearly impossible for anyone who is not a member of the "In God We Trust" mob to hold office or attain any position of authority, which requires either election or appointment.

The Constitutions of, at least, nine states explicitly prohibit the public employment of anyone who does not believe in or who expresses doubt in the existence of a "Supreme Being," which is the generic name for the Abrahamic God. Several state Constitutions including those of Arkansas, Maryland, Mississippi, South Carolina, Pennsylvania, Tennessee, and Texas, specifically state that government employees must believe in a "Supreme Being," "not deny the existence of Almighty God" or else make some similar requirement.[3] On February 11, 2009, *Bill HJR1009* was introduced to the Arkansas State Legislature in an attempt to repeal the prohibition of an atheist holding public office or testifying in court, but it died in House Committee.[4] The Supreme Court declared these laws invalid in 1961, yet they persist.

Freemasonry, with which the Federal Reserve Bank is closely associated as depicted in their art and architecture, also, requires that its male-only initiates profess a belief in "a Supreme Being." Similarly, corporations, including banks, even small, local bank branches, typically harbor an exclusionary religious culture. Monotheistic religion is critical to the success of the Holy Brotherhood's cruel endeavors. They know that those who do not go along with their religious program are likely to be heretics in other ways, as well.

Honor Among God's Thieves

As commanded by their religious doctrines, the Holy Brotherhood and their faithful Men of God band together against the infidels. They have been practicing their financial crimes for millennia, and their targets have long been witches, occultists, and pagans. Two different translations of this command to rob and dispossess witches and occultists in *Deuteronomy 18:9-14* are provided for clarity, and bolded for emphasis, as follows:

(King James Version) "When thou art come into the land which the Lord thy God giveth thee, thou shalt not learn to do after the abominations of those nations. There shall not be found among you any one that maketh his son or his daughter to pass through the fire, or that **useth divination, or an observer of times, or an enchanter, or a witch. Or a charmer, or a consulter with familiar spirits, or a wizard, or a necromancer.** For all that do these things are an abomination unto the Lord: and because of these abominations the Lord thy God doth drive them out from before thee. Thou shalt be perfect with the Lord thy God. **For these nations, which thou shalt possess, hearkened unto observers of times, and unto diviners:** but as for thee, the Lord thy God hath not suffered thee so to do."

*(The World Messianic Bible) "When you have come into the land which the LORD your God gives you, you shall not learn to imitate the abominations of those nations. There shall not be found with you anyone who makes his son or his daughter to pass through the fire, one who uses **divination, one who tells fortunes, or an enchanter, or a sorcerer, or a charmer, or someone who consults with a familiar spirit, or a wizard, or a necromancer.** For whoever does these things is an abomination to the LORD. Because of these abominations, the LORD your God drives them out from before you. You shall be blameless with the LORD your God. **For these nations that you shall dispossess listen to those who practice sorcery and to diviners;** but as for you, the LORD your God has not allowed you so to do."*

Seemingly conflicting Biblical admonitions to love and forgive, and not to steal, covet, or lust only apply among members of their brethren and their own wives and children. They do not apply to you. Instead, God instructs the faithful to rob, enslave, and kill you.

The Torah states, "The best of the gentiles: kill him; the best of snakes: smash its skull; the best of women: is filled with witchcraft."[5] Other passages of their holy books, such as *Sanhedrin 74b* and *Yebamoth 98a*, call you sub-humans and animals, over whom their God has given them dominion. One of the deadliest verses in the King James Version of the *Holy Bible, Exodus 22:18* states, "Thou shalt not suffer a witch to live." The *Koran 2:102* states that Allah condemns witchcraft, and nations dominated by that aspect of the Sinister Trinity still have secular and religious laws against witchcraft, which are brutally enforced.

This is why, in 2008, God's bankers, working with God's earthly ministers in Congress, so easily obtained U.S. taxpayer bail-outs. Congress awarded billions of U.S. taxpayers' dollars to them, even after they had already stolen vast amounts of their victims' real assets by fraud. In January 2016, they were "bailed in" by the innocents they defrauded in Greece, when they directly stole 20% of their depositors' funds.[6] When bankers confiscate the contents of your account, forge contracts, or fraudulently raise the "fixed" rate on your credit card, *there is no crime* because they are acting in obedience to God. This is why financial crimes are rampant, and the laws and courts are rigged against you and in favor of the Holy Brotherhood.

"We" is not Us

Clearly, the "We" in the statement, "In God We Trust," is not us. The "We" are the same ones who dispossessed untold thousands, possibly millions, of presumed witches in Western Europe before they brought their sadistic violence to the Americas where they did exactly the same thing again, accusing, criminalizing, and demonizing their victims while claiming real estate property and mineral wealth.

"We" means the believers, those faithful to the God of Abraham, the Men of God, the religious supremacists. The "We" are the Holy Brotherhood who administer the Federal Reserve System, the occult brotherhood of international plunderers. They "Trust" in their holy manual of accounting fraud, which has served them so well for millennia. When you pay usurious taxes, you are paying tithes, making an annual ritual sacrifice to their fictitious God.

Tithing and sacrifices are required to protect you from God's wrath and curses. While their God once demanded blood sacrifices and firstborns, now he just wants a portion, 30%, 45%, or maybe 55%, of your life force energy. If you do not go along with this scam, then God says you are robbing or cheating *him.*[7] His curses and damnation consist of a visitation from militarized law enforcement agents.

Your enemies' objective is clearly written in their own books. While they have a history of murdering witches and occultists, their main plan is not to kill you all at once, but to keep you weak, so they can ritualistically bleed you and feed upon your life force. The "In God We Trust" system, which includes the Federal Reserve System, its subsystems, and the Men of God who participate in its authoritarianism, are the present-day incarnation of our historical enemies who dispossessed and burned our foremothers at the stake. To them, this slightly less bloody scheme is a sign of their civilization's progress.

CHAPTER 4
THE MANY FRAUDULENT SCHEMES OF THE HOLY BROTHERHOOD

The Holy Brotherhood are invaders, colonizers, war profiteers, and extremely well-organized racketeers who have spent ages perfecting their repertoire of time-tested financial scams designed to covertly destroy their enemies. Below is a summary of the contents of their bag of dirty tricks:

"Let Me Hold That For You"

They promise to hold your assets in their vaults for safekeeping. Once they take custody, you may not get your property back.

Direct theft. They may simply empty your account, and you will never see your cash or gold ever again.[1]

Internal banking fraud. Bankers and their employees directly steal from account holders with impunity.

Forgery. Banks commit identity fraud in the form of forgery to open accounts and lines of credit and alter the terms of existing contracts. Lenders cooperate with external identity thieves to open loans and credit cards in their victims' names.

Refusal to allow the withdrawal of funds. If you question fraudulent activity, they may defensively confiscate your entire account.[2] Meanwhile, your transactions are being watched, recorded, and judged under Know Your Customer (KYC) and Anti-money Laundering (AML) regulations to determine if *you* are a criminal. If a bank teller deems you "suspicious," they may freeze your account, confiscate the contents, and send cops to your door.

Fees: Bankers charge confiscatory fees for any reason they dream up. There are overdraft fees, ATM fees, maintenance fees, returned deposit fees, lost or stolen card fees, fees for paper statements, inactivity fees, excessive activity fees, telephone bank transaction fees, Internet banking fees, branch banking fees for transacting with a human teller, fees for making a late payment, and fees for exceeding your credit limit.

Outrageously high, hidden fees are incorporated into the cost of products and services you buy from any seller with a merchant account. For instance, when you buy groceries, even when you pay cash, a large percentage of the price goes to cover the costs of the fees the merchant must pay the banks to process bank cards. This is why bankers eagerly offer customers "free" debit cards and interest-free credit cards. But, they are far from free in every respect.

Currency exchanges charge fees to change one currency for another. Some custodial exchanges charge a substantial withdrawal fee, too. Since they do not always adhere to a standard exchange rate, some get a hidden "fee" by using alternate rates, calculated in their favor.

Usury

Usury is the ancient scam of not only collecting interest, but setting up the borrower to fail so bankers can take an asset. They continue to innovate on this fundamental fraud.

Creative mathematics. They apply a variety of devious, standardized practices for calculating interest, payments, and other terms of loans, which always work in their favor. For instance, they apply monthly payments to interest before the principle sum or apportion the largest part of monthly payments to interest, which keeps the borrower locked into the loan.

Deceptive accounting practices. Centuries ago, the accounting guilds established misleading practices and terminology to purposely confound those outside the profession. Double-entry accounting, in which every transaction is recorded twice, once as a "debit" and once as a "credit," is an example of this. It was first described in the appendix of a book, *The Book on the Art of Trading (Libro de l'Arte de la Mercatura)*, by Benedikt Kotruljević̇in in 1458. It may be used as an honest bookkeeping practice, however, it is used by bankers as a tool of deception and fraud.

By means of double-entry accounting, a bank's money is seemingly created out of thin air, then used as consideration in making loans. In 1968, this fraud was famously exposed in a court of law by the *Credit*

River Case. According to the memorandum of this case:

Plaintiff admitted that it, in combination with the Federal Reserve Bank of Minneapolis, which are for all practical purposes, because of their interlocking activity and practices, and both being Banking Institutions Incorporated under the Laws of the United States, are in the Law to be treated as one and the same Bank, did create the entire $14,000.00 in money or credit upon its own books by bookkeeping entry. That this was the Consideration used to support the Note dated May 8, 1964 and the Mortgage of the same date. The money and credit first came into existence when they created it. Mr. Morgan admitted that no United States Law Statute existed which gave him the right to do this. A lawful consideration must exist and be tendered to support the Note. See Ansheuser-Busch Brewing Company v. Emma Mason, 44 Minn. 318, 46 N.W. 558. "The Jury found that there was no consideration and I agree. Only God can create something of value out of nothing."[3]

This is how most, if not all, loans are made. The bankers risk nothing because the money they pretend to lend is fictional. When the borrower defaults, they do not "take back" property, as many people suppose, rather they *acquire it.*

Banks simultaneously regard loans as liabilities and assets, which they package and trade as securities. "Cooking the books" is built into this accounting system. The innovation of double-entry accounting combined with usurious fractional reserve lending forms the foundation for other fraudulent schemes of the central banks.

Central Banking

By the issuance of usury-based currency to governments, internationally-owned central banks disguise their usury as taxation. The Holy Brotherhood has a tentacle in every country, sucking the life's blood of innocents, while amassing the largest fortune in the world.

Inflation. When the bankers create air money, the nation's debt increases and the currency loses some of its buying power. The cost to purchase goods and services goes up and the value of real assets go down. Once a nation's economy is sufficiently wrecked, the Holy Brotherhood's members buy what is left, such as houses, land, and mineral resources, at bargain prices.

Fomenting wars. The central bankers foment wars and conflicts, then lend money to all the parties involved at usurious rates. They own the corporate media and the manufacturers, which create the propaganda and the tools of war. War no longer seems senseless once you understand this ancient, holy scam.

Organized crime and terrorism. The banks work with the mafia and terrorists, including notorious Mexican gangs and Al Qaeda.[4] According to Antonio Costa, a former head of the UN Office on Drugs and Crime, many banks may have collapsed during the 2008 economic crisis if not for the liquidity provided to them by loans they made to mobsters involved in trafficking and terrorism.[5]

Rigging the Game

Creating regulations, then skirting them. The Holy Brotherhood design an economic crisis, then once the scheme climaxes, they are among the financial experts who propose more regulations with loopholes, which they quickly weasel their way through to resume their fraud. For instance, after the economic catastrophe of 2008, new consumer protection regulations were devised, but the bankers continued issuing subprime, "Class D" mortgages through third-party corporations, which are not technically banks and not subject to banking regulation.[6] When regulations forbade them from issuing second mortgages as down payments, they just renamed them "personal loans" and proceeded. Such fraud, even when perpetrated on a grand scale, carries no criminal punishment for God's elite, who are doing his good work.[7]

Securities rigging. Stock markets and commodities trading are every bit as much gambling as spinning a roulette wheel. The game is rigged in favor of the house, which is owned by mobsters. Similarly, every stock market crash or securities scandal ever experienced by the United States has been rigged by insiders. Insider trading is not illegal if you are a member of the Holy Brotherhood.

Coercion

A central bank, once implanted, is like a bullet lodged close to the heart of a nation. It needs to be removed for the health of the body, but doing so is mortally dangerous. This is why it is difficult to get rid of a central bank. They hold a country hostage, making ever greater demands, threatening even worse economic disaster or military force to get their way.

Force. Any place a central bank exists, individuals are forced to use it by the threat of imprisonment and confiscation of property. Their members infiltrate the government and they bribe and extort government agents in order to obtain what they want, whether it is charters, licenses, favorable laws, impunity, bailouts, or bail-ins.

Psychological warfare. They gain the public's compliance by convincing them, by means of both religious and secular propaganda, that paying a tribute ("tributum" is Latin for "tax") to the God of an international criminal cartel is patriotic.

Early examples of secular propaganda include, *The Jack Benny Program,"*The Income Tax Show," which aired October 16, 1964; *The Beverly Hillbillies,* "Jed Pays His Income Tax," which aired April 3, 1963; and *The George Burns and Gracie Allen Show*, "The Income Tax Man," which aired March 1, 1951. Despite the existence of the Federal Reserve System since 1913, most Americans did not pay income tax before the 1950s. These television programs were designed to get the public used to the idea of illegal, quasi-government intrusions on personal privacy by using humor and trusted, familiar faces on television. They normalized this activity, presented it as wholesome and all-American, and disseminated other disinformation about the organization.

The color of office. What God's bandits once accomplished through open plunder and massacre, they now achieve by pretending to act under the legitimate authority of a government office.

Data collection. In an exercise of total control, they collect personal and financial data on everyone who has a Social Security Number, including infants. Some of the data is incorrect and defamatory, but it is shared, sold, used by lenders, intelligence firms, private investigators, potential employers and landlords, and made available on the black market of the Dark Web.

Faith and trust in God. They fused faith in God with faith in government. Most people, whether religious or not, entrust every aspect of their lives to God's earthly ministers.

The Worm Turns

> "Treade a worme on the tayle, and it must turne agayne."
> — William Shakespeare, Henry VI, Part 3.

Now, the Holy Brotherhood's abusive "In God We Trust" system is experiencing resistance. This banking heresy is made possible by the worldwide growth in information technology, massive Internet participation, very fast computers, and sophisticated digital encryption. In addition, the astrological timing of it could not be more fortuitous.

CHAPTER 5
BANKING IN THE AGE OF AQUARIUS

In astrology, an age is about 2,160 years. The story of Jesus Christ, fisher of men, is an astrological allegory of the birth of the Age of Pisces, a time in which the world came to be almost completely dominated by religions of hatred and war disguised as love and peace. The Piscean Age began at approximately the first *anno domini*, which means "Year of Our Lord." The term, "lord" means "ruler," "master," "commander," or "one who dominates." According to astrologers, the Age of Pisces is characterized by religion, authoritarianism, illusion, deception, fear, fraud, imprisonment, and the hoarding and hiding of important information.

Occultists have been expectantly awaiting the "golden dawn" of the new Age of Aquarius, since the nineteenth century. While the world is still on the cusp between these two ages, the Piscean influences continuously weaken, as the influences of the new Aquarian Age become stronger every minute. While the planetary influences of the Piscean Age remain, its masters frantically tighten their stranglehold. In due time, if humanity prevails, they will not entirely vanish, but they will be diminished and unable to return to power for a long time. With them will go the tyranny of centralized systems of hierarchical authority, in which a very few lord the power of life and death over the great many.

Domination in the Piscean Age

The Holy Brotherhood came to dominate most of the world during the Age of Pisces. They epitomize this terrible age because they made literal prisons of many parts of the world and engendered even more restrictive virtual prisons in the minds of the people they colonized.

Wherever they went, the Holy Brotherhood employed the same methods. They robbed, enslaved, tortured, terrorized, and murdered the indigenous pagans. They replaced our ancestors' wisdom with their fictional God and made him the authority on everything including us. They excised and revised human history to suit their nefarious purposes, supplanting our culture, science, medicine, society, and money with their own perverse versions of these things.

They forced on and reinforced in the people a love for God, the authority of which flows from him to his earthly ministers. Their obliteration of witches and witchcraft from the cache of limited information permitted to the public is now so complete that they sometimes claim we do not exist and, in an effort to cover up the true nature of their crimes against us, that we never have.

By persecuting and murdering witches, they restricted the ability of the civilized masses to properly see, hear, feel, and intuit their surroundings, rendering them able to perceive only their masters' illusions. They hobbled them by denying inherent abilities, which were taken for granted before the arrival of the Holy Brotherhood. They have impeded communication by limiting, twisting, stretching, and inverting the language. They have exercised complete control over the minds of the masses by erecting gatekeepers at every possible avenue of escape, which appear to offer liberation, but only serve to herd the errant prisoner back into another virtual prison cell.

They have long ruled by permeating every aspect of life with their hierarchical organizations. They continue their crimes against us by means of their occult banking system, the keystone of their overarching prison system, which would crumble to dust if it were plucked out.

The Luciferian Age of Aquarius

The Age of Aquarius is the time when the planetary frequency harmonic is right to bring an end to the reign of the dark overlords. It is a time of upheaval, in which the planet undergoes physical changes and people, especially those in harmony with the Aquarian Age, experience paradigm shifts as many hidden things are revealed to all who dare to look. Right now, the prison door that slammed shut at the beginning of the Age of Pisces remains locked, however, a window has appeared and an unconventional, new avenue of escape presents itself.

As the influences of the old age wane, the Holy Brotherhood's illusion of power is ebbing. Their edifices are crumbling. Those sincere in their faith in God are panicked, fearing the Apocalypse. They are "prepping" for doomsday, and some are literally fleeing to the hills because what is inevitably coming is the end of *them*. Their own holy books, based on the occult science of astrology, predict it and while they may not believe in astrology, they always believe "God" without question.

Meanwhile, the Holy Brotherhood are running their propaganda machine at full volume, silencing, censoring, and defaming heretics, just as they have done for centuries. The international bankers hold tense meetings, trying to decide what to do about all this rebellious *evil*.[1] Many times in the past several years, they have convened to discuss how to reckon with the Antichrist who has come to do them in.[2]

It is accurate to call this new age the "Information Age" because, in many ways, they are one and the same. To the Holy Brotherhood, it is an "Age of Evil" The word, "evil," is an inversion in which the acquisition of knowledge, which is true information, represents an offense to the God of the Holy Brotherhood. To be "evil" is to be "like Eve." From a witch's perspective, Eve represents the perfection of Goodness, because she defied our ultimate enemy, God the Fraud. Therefore, the Aquarian Age is, also, the Age of Eve, who took the fruit of the Tree of Knowledge and shared it, upon which God opened the prison gates. Thus, learning the truth about the illusion of God's authority set the prisoners in the garden free.

Judeo-Christian Apocalyptic authors, who acknowledge the Age of Aquarius, commonly write of their fears of Lucifer and Illuminism, which are concepts associated with this age, in which the light of Lucifer, the Morning Star, also, called "Venus," and "Astarte," shines down from the heavens, destroying the shadowy illusions of God and religion, revealing important truths. While Christian authors acknowledge the existence of the same pernicious system, they are unable or unwilling to accurately name the forces behind it, although they correctly equate rebellion against God with Lucifer and being Eve-like, or evil. Lucifer is, also, associated with the legend of Prometheus, who brought the technology of fire to humans against God's will. As Prometheus, Lucifer represents the threat of information technology to the power of the Holy Brotherhood and the Men of God.

The world is not ending, only the Piscean Age and the mass delusions of the faithful, who are in a mental prison. Just like prisoners in actual prisons who have been confined for very long, they fear what awaits on the outside. Fortunately, their fears of the Luciferian Age of Aquarius are fully justified.

A New Possibility

Bitcoin is the metaphorical window in the prison cell, illuminated by the first rays of a new dawn. It is a brainchild of the Age of Aquarius and, while it is not the be-all and end-all, it is a very good start. It is to the banking system what *The Hermetic Order of the Golden Dawn* was to the free masonic lodges. Just as *The Golden Dawn* revealed the inner workings of occult lodges and made the information irrevocably public, so Bitcoin reveals information and tools to undermine the secrecy and authority of the financial cartel. Its creators understood the structure of the occult banking system and designed Bitcoin to smash it to smithereens.

The possibility at hand is the true liberation of the mind, however, for you to fully avail yourself of it requires a paradigm shift because it is a technology and a philosophy very different from that of the prevailing order. Upon first learning about it, most people need a few days to fully assimilate the information and absorb its impact. Moreover, because they have learned to profit by Bitcoin's misuse, but suffer when it is used correctly, the Holy Brotherhood continue to issue a barrage of disinformation about it through corporate media and government bureaus, which may create obstacles for those who faithfully accept the authority of God's earthly ministers.

Right now, the light of Lucifer is shining down on the earth ever brighter, gradually driving away the last shadows in which the overlords of the Piscean Age reside. But, the Age of Aquarius is only a chance at liberation and if humanity does not prevail against the masters of the old age, the world will fall into a prison of even darker despair.

The world may very well fail because it takes bravery to recognize that you are in a prison. Then, once you recognize this, you must be able to perceive the beam of light streaming through the tiny crack in the wall, there discern the shape of a window, and dare to use it to step outside.

Many people cannot see that the human race is in bondage. If they cannot see this, they will maintain the illusions of the old age and continue to suffer as a result. But witches and occultists perceive the world very differently from most other people and because of this are well-suited to making the necessary mental shifts. Many have already made this transition.

The current system is simply a mass illusion, albeit a pervasive one. It is predicated on the authority of God. But, the Holy Brotherhood, their fictional God, and his secular ministers have no power over witches, who are heretics by nature. Long ago, we devised ways of dealing with our most formidable enemies, with which this new information technology fits perfectly. But, before explaining how the Holy Brotherhood's authority may be immediately circumvented and eventually overturned, it is important to more closely examine the architecture of our historical enemy's most vital system.

CHAPTER 6
CENTRAL BANKS AND OTHER SECRET SOCIETIES

To understand the pivotal component in the Holy Brotherhood's mechanism of domination, it is necessary to consider the organizational structure of a corporate central bank, which is like that of any secret society. Both are formed in the shape of a conic pyramid, in which circles of power and knowledge run from the base of the structure to the top, and from the outer surface to the inner core.

This architecture enables secrecy and the concentration of power in the hands of a single authority, at the center and top, who is the only one who has full information about the organization. How the organization appears to those who operate at the bottom and outside of it is very different from how it actually is because their information is limited or they are provided with disinformation. At each lower rung, they obey their masters because they have faith and trust in authority and receive rewards for their loyalty. The most corruptible in the organization are permitted access to the inner circle and rise to the top.

Corporations, governments, bureaus, most religions, the military, the educational system, the medical system, and so on, are constructed the same way, with centralized, authoritarian control coming from the top in a hierarchy of power and knowledge. From birth to death, this concept of authoritarian, centralized control pervades the minds of most people. Our entire modern civilization, even book clubs and Tupperware parties, are all run on the same model. This old, authoritarian paradigm is the Holy Brotherhood's ruinous gift to all the heathens they conquered during the Piscean Age.

What is a Secret Society?

A secret society is one that conceals itself in specific ways. By the narrowest definition, it is a lodge that has prescribed rituals and an oath of secrecy, which is secured by explicit threats. But, not every secret society is a lodge. While a large number of secret societies exist, whose names are unknown to the public, especially if tiny, independent covens of witches are taken into account, many exist that are known by name, although they may conceal other information about themselves, such as the identities of their highest ranking members, the location of their headquarters or meetings, and their purpose.

A few well-known examples of secret societies include the Rosicrucians, the Francis Bacon Society, the Order of the Friars of St. Francis of Wycombe (the Hellfire Club), Greek fraternities, the Church of Jesus Christ of Latter-day Saints (The Mormons), and the Freemasons. Some secret societies are benevolent and work for the spiritual progress of mankind or to further a secular philosophy and may not have an ulterior motive. But, a majority of known secret societies exist to obtain unearned advantages for their members, who are usually male. Most notoriously, Freemasons and Greek fraternities install their members in positions of government and corporate power. Secret societies usually serve a combination of purposes, including to obtain career advantages, political favors, leniency from law enforcement agents and judges, and to hide unethical and criminal behavior.

Some, such as the Mormons, are a corporate commercial enterprise in the guise of yet another charitable Christian denomination. It is an example of an extremely authoritarian, hierarchical, and secretive society. Just inside the shell of the structure, there are Freemasonic rituals, signs, and tokens or symbols. Unearned life advantages are provided to its male members who are all priests of some order and rank. The priesthood orders, range in power and authority from the bottom to the top: The Aaronic; the Melchizedek; and the Patriarchal. Within each of these three levels are sub-levels: Deacon; teacher; priest; bishop; presiding bishop; elder; high priest; seventy; patriarch; presiding patriarch; and apostle. The priesthood is only open to men and boys of 16-years of age. They hold their meetings secretly, apart from the women and girls, who are subordinates to even the lowest priesthood holders and whose duty is to mind children and be groomed to serve the men. Men who rise to the level of the bishops over wards may be tapped for membership in yet another secret society, which is known as the Second Anointing,[1] in which they believe themselves transformed into earthly gods with the power to commit violent crimes without fear of afterlife punishments.

The Mormons even rank the after life from the bottom to the top as the Telestial, Terrestrial, and Celestial Kingdoms, with each of these major divisions containing a hierarchical ranking of subdivisions of privilege and authority. At the inner circles, members take an oath of secrecy that until, recently, included the threat of death and loss of possessions. They have their own intelligence division, called "The Strengthening the Church Members Committee," which spies on suspected apostates and ex-members who talk too much. It is the modern incarnation of the murderous nineteenth-century military fraternity, the Danites, also, known as Destroying Angels. The whole of the organization is held together by faith and trust in a hierarchical system of earthly authorities who claim to speak to God, with the presiding president regarded as a living prophet and the ultimate authority on God's capricious will. It is public information that the Church of Jesus Christ of Latter-day Saints is comprised of a conglomerate of numerous Limited Liability Companies with diverse investment portfolios,[2] and its institutionalized sexual abuse of women and children is a poorly kept secret,[3] however, it is likely that only the president and his immediate counselors know the true mission of this international secret society.

Commonly, secret societies hide their true objective behind a facade of ordinary commercial or charitable activity. Some are secret societies within a secret society. Characteristically, powerful, established secret societies dislike competition of any kind and regard members of other secret societies as enemies, since many such societies undermine the power of entrenched authorities.

The Catholic Church as a Secret Society

The Roman Catholic Church is an example of a well-known secret society, which appears to be a religious organization that promotes faith in God and performs charity work. In fact, it is an international commercial enterprise in possession of its own military orders, established during the crusades, and its own private bank, The Istituto per le Opere di Religione (IOR or Institute for the Works of Religion),[4] which is not part of the central banking system and has been entrenched in scandals and litigation relating to fraud and embezzlement since its inception.[5] The whole of the Catholic Church is comprised of a hierarchical network of support organizations or subsystems, each performing different functions. Many of these subordinate organizations are individual corporations. For example, in the United States, an archdiocese may be registered as a charitable, tax-exempt corporation.

At the bottom and outermost realm of power and knowledge, rank the general members, the parishioners, whose functions and level of knowledge about the organization are most limited and compartmentalized. Their duty is to trust the authority of the priests in their respective parish, to protect him, even when he is involved in the most heinous atrocities, to obey, to selflessly serve him, and to make tithes and offerings. The masses at the bottom are equivalent to "porch masons," standing on the periphery, participating, feeding the body of the organization, but of low rank and intentionally provided with misinformation by those in authority. As payment for their fidelity, some members receive tangible benefits from their membership, such as business connections, however, they know little about the organization to which they belong.

Above the general membership is a succession of ranks, each with its own hierarchy and compartmentalized with different degrees of knowledge and authority. They rank from the lowest to the highest, as follows: Priests; bishops; cardinals; and, finally, the pope. Members of the highest, innermost circle of the papacy conduct business that is unknown to the bishops, priests, and parishioners. At the very top of this pyramid, removed from the physical world, is their God, three personages in one, who is the invisible head of the organization, whom the pope is literally supposed to embody, at least, when exercising his authority as head of the organization. Thus, the pope is the ultimate authority, God in the flesh.

As a monolith, the Church may be envisioned as a conical, grand pyramid comprised of many sub-pyramids, all working together like a well-oiled machine. Examples of sub-pyramids include the Benedictines, the Jesuits, Opus Dei, the Knights of Malta, the Swiss Guard, and, of course, the Vatican Bank.

Faith and trust in God and his earthly embodiment, the Pope, are the cogs that unify and animate the entire body of the mechanism. At each level of the hierarchy, the members hold their masters in reverent awe. For the lowest member of a parish, obedience to the lowest-ranking and debauched priest is obedience to God.

Corporations

Corporations, including banks, are bodies created by men based on legal statutes. They are homunculi, bodies without souls, endowed with the legal power to act like persons, but are not persons. Like the alchemists' creation, a corporation can be used as a servant, a decoy, or a scapegoat, by means of which its creator can act without being held accountable. For instance, it is possible for the artificial, corporeal body to be sued or go bankrupt without its master being harmed. Corporations are shields against personal liens, liability, and usurious taxation. Privacy corporations and offshore corporations may conceal the identity of an individual or group. Sometimes they transact business by the issuance of such financial instruments as bearer's bonds, or they devise some paper-shuffling scheme to hide criminal activity. They may allocate certain functions to legal firms, through which the law permits secret activities.

There are many different types of corporations, which include those that are for-profit and not-for profit, and those that are publicly traded and those that are not. The organization of a corporation and the protections endowed to it are dictated by lawmakers and the IRS. The latter determines how they will tax such a body, so that the Holy Brotherhood and the Men of God do not pay usurious taxes, while their enemies pay and pay some more.

Government entities charter, or grant licenses, for a special purpose, to some corporations. Such is the case with Freddie Mac and Fannie Mae, two chartered U.S. corporations that guarantee loans. Other common chartered corporations include cities and villages within states.

While corporations can provide legal protections, even a high level of privacy, when formed in the proper jurisdiction, they are not always beneficial to the one who establishes them, especially once they offer public shares. Commonly, they are a means for parasitic personalities to wrest control away from creators and inventors. The shareholders ultimately own the company and based on the number and class of shares they hold (some shares hold more weight than others in a vote) decide its fate. In this way the creator may be deprived of the right to the fruits of his or her labor by those in a position to do so. Since any business endeavor that grows to any size is subject to greater tax liability, it is essentially forced by the current system to establish a publicly traded corporation in order to defer the burdens placed on it by the system.

Some tax-free, non-profit corporations, which are trusts or endowments, exist both as tax dodges and mechanisms of psychological warfare. These are not the same as public charities that collect donations for the poor. While such corporations claim to be philanthropic, charitable organizations, "doing good" for humanity, they covertly manipulate society, defining medicine, science, the arts, history, and education as they see fit. They shape their enemies' opinions, and desires. The Rockefeller Foundation and the Ford Foundation are two of the earliest and largest such organizations to rise up about the time of the formation of the Federal Reserve System.[6]

In the United States, corporations bear a centralized, authoritarian structure with the Chief Executive Officer at the top and core. Beneath him are management, a compartmentalized hierarchy of employees, a board of directors, which ostensibly exists to guard shareholders' interests, and the shareholders, who, despite being the organization's financial owners, are usually the least informed about what is happening in the company. They are often purposely deceived by the withholding of critical information about the state of the company or by deceptive practices that artificially inflate the value of the company's stocks.

Those in the middle levels protect those in the upper levels, either out of ignorance or a sense of self-preservation, from those at the lower levels or from outsiders who might obtain secrets about the corporation. Corporations sometimes require oaths of secrecy in the form of Employee Non-disclosure Agreements. Even when this is not a requirement, they reward loyalty and punish whistleblowers by firing and blackballing them. These bodies without souls develop their own toxic personalities because they tend to promote people who have tendencies and values similar to that of the leader and his core group and weed out dissenters, a phenomenon called "corporate culture."

Because of its centralized nature, a corporation that serves a benevolent purpose can be easily subverted, especially when the individual at its head retires or dies. Sometimes when a corporate head dies, so does the corporation. Corporate structures are, also, subject to corporate laws. For example, if a government entity subpoenas information, then it must be provided by the corporation under severe penalties to its CEO. In this way, corporations are vulnerable to attacks by corrupt governments. Since they, also, centralize customer data, they are vulnerable to leaks and hacks. The secretive nature of corporations enables fraud, which is why government oversights are deemed necessary, although they are always insufficient.

Banks as Secret Societies

A bank is a corporation whose business is to issue, hold, lend, exchange and transmit funds. They take advantage of the secrecy their corporate structure affords. Despite their friendly facade, banks do not have a benevolent nature at their core and, in the United States and most other countries, there is not a single one that is not directly connected to the central bank. Credit Unions and payment processors are all connected to the banking system, too. Because they deal with the funds of individuals and organizations, they wield a dangerous level of power, which they regularly abuse.

Banking is at the heart of a body comprised of many similarly designed, interrelated structures. Prominent bankers are involved in corporations that specialize in insurance, advertising, traditional publishing, movies, television, radio, and other communication industries. The contents of the bankers' ledgers are secrets known only to them. Furthermore, the identities of the men at the top and core of the international banking network are secret.

"In God We Trust"

Banking began in temples and some banks, such as the Knights Templar, were militarized, religious secret societies. All banks rely on the trust and faith of their customers and, when this fails, they rely on force. As a result of the Holy Brotherhood's centuries of intergenerational religious programming, the mentally colonized public is vulnerable to trusting highly untrustworthy third parties. Their trust in centralized, top-down authoritarian structures is so culturally entrenched that it appears natural and, therefore, passes unquestioned.

As his earthly ministers, the bankers exercise God's worldly judgment on you. They decide if you will be allowed the privilege of paying them usurious interest in order to obtain goods at inflated prices, which they control. They decide whether or not you are worthy to have a personal bank account or a merchant's account, who you can and cannot do business with, and what you can and cannot buy or sell. They punish their enemies by denial of economic participation.

To maintain their God-like authority, they promote a belief in a higher power. Faith and trust in God holds their many subsystems together as one monolith, as one fraud under God, indivisible. The "In God We Trust" system is intentionally designed to facilitate fraud.

Monopolization

It is the Holy Brotherhood's nature to gather and coalesce many things into one, the same way they assembled many regional gods into one monotheistic, all-powerful, all-seeing God. They have conditioned the masses to do the same and to seek out a singular authority in every matter, whether religious or secular. Even the original, unofficial U.S. motto, "E pluribus unum," appears to be theirs. It may be interpreted as, "From many, one." Universality, unity, oneness, uniformity, conformity, communality, compliance, monopolization, monotheism, and submission are all ideas associated with the Holy Brotherhood.

The central bank and its international network of private banks comprise a universal monopoly, maintained by secrecy and force. Even those who have long been aware of this nefarious, international secret society of financiers are forced to use their phony money. At least, that was the case until very recently.

CHAPTER 7
WHAT ARE BITCOIN AND THE BLOCKCHAIN?

Bitcoin is the intrinsic transaction medium of the Internet, the offspring of the Information Age, born at the height of the 2008 banker-induced economic crisis. It is the natural result of the modern state of information technology and the increasing need for privacy, especially financial privacy. Its decentralized, non-proprietary model is the antithesis of a corporation with the built-in potential to diffuse the power of the traditional banking system.

While Bitcoin is the first decentralized, peer-to-peer cryptocurrency based on free, open source software code, it is not necessarily the best one as there are already many innovations, some of which better represent the mission of Bitcoin. But, it is important to understand Bitcoin and its philosophy before moving on to the subject of its improvements.

The Origins of Bitcoin and the Blockchain

The first conception of Bitcoin is recorded in 2008 in a post on the *Metzger, Dowdeswell & Co. LLC Cryptography Mailing List* at metzdowd.com. It was first detailed the same year in a white paper by a pseudonymous entity, Satoshi Nakamoto, entitled *Bitcoin a Peer-to-Peer Electronic Cash System.*[1] Basically, Bitcoin is Internet currency that functions like digital cash, allowing one-to-one transactions without a middleman.

Bitcoin and the blockchain are based on binary digital ("bit") mathematics. The blockchain is the digital ledger that records Bitcoin transactions. The blockchain is based on free, open source software code, which anyone may audit, replicate, or edit. Bitcoin's blockchain ledger is distributed worldwide across a large number of servers and available to all users at any time. This public ledger allows for a transparent, chronological, indelible record of accounts and transactions, and contains other features designed to discourage fraud and encourage honesty. By means of the blockchain, any user can transact directly, peer-to-peer (P2P), person-to-person, with any other user, with no need for a trusted third party.

Open-source software (OSS) is computer code released with a copyright that conveys the legal right to anyone to study, change, and distribute it for any purpose. Since the blockchain is infinitely replicable open source software code, owned by everyone and no one, an infinite number of different digital currencies, called "altcoins," may be created with it.

The replicability of the blockchain is often cited as a concern by those who believe Bitcoin is a commodity to be traded on the stock market, but it is an important feature to those who understand that Bitcoin is a medium of transaction because it provides you, the banking heretic, with infinite choices of competitors or "altcoins." By design, Bitcoin is not singular nor supreme among cryptocurrencies. This ensures that if Bitcoin becomes manipulated by the Holy Brotherhood, as it has been, there is always another option.

Bitcoin has only recently appeared in the mainstream media, prompting God's faithful to fear that it is the "Mark of the Beast," however, it is merely the first digital currency to gain any traction.[2] Its flat, distributed, peer-to-peer structure is the result of lessons learned from the demise of past digital payment systems, in particular, Ecash by DigiCash, Inc., a centralized, cryptographic digital currency corporation. It existed from 1989 until 1998 when it went bankrupt.[3] It was, also, intended to thwart "Big Brother."[4]

Other aspects of the Bitcoin blockchain were foreshadowed in an article written by Wei Dai in 1998, entitled "B-Money."[5] Dai wrote about community-updated, public ledgers, proof of work, financial rewards for processing transactions, lack of central authority, and contracts validated by means of digital signatures, all of which are concepts incorporated into the blockchain.

Bitcoin's philosophy is rooted in the Cypherpunk movement, which began in the 1980s and advocates the use of strong encryption to facilitate liberty through privacy. Cypherpunk philosophy is exemplified by a 1993 article by Eric Hughes, entitled "A Cypherpunk's Manifesto," which states that "privacy in an open society requires anonymous transaction systems," and acknowledges that "faceless" governments and corporations cannot be expected to grant people privacy out of the goodness of their hearts.[6] Additionally, Wei points out that two parties to a transaction should not have any more knowledge about each other than "that which is directly necessary for that transaction."[7]

Bitcoin Versus Banking

Bitcoin is more than just a way to send or receive a payment. It is a philosophy of liberty through personal and financial privacy, achieved by means of strong encryption. It is the complete antithesis of the "In God We Trust" system. It facilitates "trustless" transactions, removing its users from thousands of years of religious authority. Its power to undo the Holy Brotherhood lies in its faithlessness and infidelity to their ultimate authority, God. Most importantly, *Bitcoin is banking heresy.*

Bitcoin represents a different business model from the corporate structure developed over the past few centuries. Its design is an inversion of the monopolistic structure of the traditional banking system. By contrast to the well-organized, authoritarian, conical pyramid, Bitcoin's structure is flat, egalitarian, and purposefully chaotic. There is no Bitcoin corporation and no one owns the blockchain technology. It is free and open to everyone.

While the bankers' ledgers are kept private and hidden, Bitcoin's are public and available for inspection by anyone at any time. Bitcoin is designed to remove the opportunity to commit fraud and forgery, which are endemic in traditional banking. While the traditional banking system encourages and rewards insider corruption, incentives to honesty, for those who "mine," or process Bitcoin transactions, are built into the blockchain.

Unlike the U.S. dollar and other central bank-issued currency, Bitcoin is not a debt note. While the traditional banking system is built on usury, Bitcoin is generated by commercial activity, computational energy, and electricity. The Federal Reserve Bank can flood the economy with an unlimited amount of U.S. currency, which causes its devaluation and creates inflation. By contrast, Bitcoin is built on an anti-inflationary model. A fixed amount of Bitcoin is created every ten minutes and the number of Bitcoins is capped at 21 million.

There is no corporation, no owner, no CEO, no centralized database, no accounting department, and no one to whom the Holy Brotherhood's inquisitors can issue a subpoena or warrant. No one else has access to your account. Moreover, no proof of identity, no driver's license, no Social Security Number, and no credit or background checks are needed to acquire an account, therefore, identity theft and identity fraud are impossible. No authority whatsoever is in charge. Because of its globally distributed nature, in which the digital information comprising the Bitcoin ledger is stored across a large number of servers outside, there is no central point of failure. Therefore, it cannot go bankrupt nor be shut down.

No one on the inside of Bitcoin can cook the books because there is no inside. There is no one at the core because there is no core. There is no one at the top because there is no top. There is no teller, no accountant, no broker, no branch manager, no fraud department, no regulator to oversee or interfere with it, and forgery is impossible.

Bitcoin Does Not Observe the Sabbath

Bitcoin operates locally and internationally 24-hours per day, 7 days per week, with no weekends and no holidays. You can transact business freely anywhere, with anyone, at any time, for any purpose whatsoever, without waiting for a third-party to open its doors.

No Permission is Required

While the worldwide central banking system is a hierarchical, centralized monolith, Bitcoin is equitable and scattered to the winds, thus neutralizing the possibility of authoritarian control. The banking system functions on the need for authority, tracking and identification of account holders, oversights, and government regulations, all of which favor the Holy Brotherhood. In contrast, Bitcoin operates with no faith or trust in authority. It functions peer-to-peer, person-to-person, without a third-party to interfere in the transaction.

There is no authority, no one to apply to for permission, and no one to deny you access to Bitcoin for any reason, simply because the possibility for anyone to presume such authority does not exist within the system. There is no God in Bitcoin.

Fair and Sometimes No Remittance Fees

Traditional banking involves all kinds of expensive fees. For instance, their money wiring services are slow and remittance fees may run from 5 to 25%. By contrast, you can quickly send Bitcoin anywhere in the world for a low or no fee. While there are no hidden or confiscatory fees in

Bitcoin, sometimes there are "mining fees," which cover the cost of processing transactions. These fees go to individual miners on the blockchain, who operate specialized computer hardware for this purpose, which requires a personal investment and the expense of electricity. Anyone can participate in Bitcoin mining. The sender can set the amount of the mining fee to speed processing and there are no restrictions on the amount of the Bitcoin transaction.

How Bitcoin Defies Regulation and Seizure

Because Bitcoin is not issued by a central bank or a government, it is not theirs and is out of their control. It remains so as long as you keep it out of the hands of centralized, trusted third parties. Third parties, such as custodial cryptocurrency exchanges, are subject to government regulation and vulnerable to warrants, subpoenas, hackers, and internal fraud. Bitcoin is meant to be used peer-to-peer with no third party. It is based on privacy, so as long as you keep your account's decryption key code (your private key) secret, no one else can get the contents of your account.

Banks are the quintessential trusted third parties. When you deposit cash in a bank, it is the bank's property and they only give you a promise that they will give it back to you. People have difficulty getting their money out of bank accounts for a variety of reasons, such as IRS seizure, a court judgment, flagging and investigation, or a bank employee simply deems you "suspicious."

If you deposit or withdraw "suspicious" amounts of cash, your bank account is subject to investigation and seizure under "structuring" laws, according to the *Bank Secrecy Act of 1970*.[8] Your account is seized without due process, and you will likely never get it back. By contrast, if sums of cash are converted into cryptocurrency, they may be held in safekeeping, in the possession of the owner and without fear of seizure under fraudulent pretexts.[9]

Similarly, Bitcoin cannot be seized by the government in the event you become subject to some judgment or lien placed on you or your property by the courts. Nor can police, engaging in civil asset forfeiture, steal Bitcoin from motorists. In some states, law enforcement agents pull motorists over along the roadside and pocket jewelry, cash, and other valuables, and siphon off debit card and prepaid card balances using ERAD (Electronic Recovery and Access to Data) machines.[10] In addition, Bitcoin eliminates the need for offshore accounts to hold liquid assets.

Despite the fact that Bitcoin is merely digital codes stored across a large core of servers around the world, there have been efforts by lawmakers in the U.S to regulate Bitcoin as if it were an object you can carry in your pocket. An illustration of the futility of government regulation and seizure of properly maintained Bitcoin may be found in a proposed senate bill, *S.1241. Combating Money Laundering, Terrorist Financing, and Counterfeiting Act of 2017, 115th Congress (2017-2018),* which criminalizes carrying $10,000 in Bitcoin across a U.S. border.[11] Even if a law like this one were passed, it would be pointless since Bitcoin is not tangible and it is not necessary to travel with your account information in order to access it from almost anywhere in the world.

How Bitcoin Eliminates Identity Theft and Banking Fraud

Identity theft is a prevalent and rarely prosecuted crime. It is, also, one in which, as the victim, you are considered guilty until you prove yourself innocent to the Holy Brotherhood and their Bishops, which is time-consuming and often impossible. The crime of identity theft exists because the Social Security Number, the Holy Bankers of the Federal Reserve System, their Holy Bishopric of credit reporting agencies, and priesthood of usurious lenders exist.

In the United States, each individual, including newly born infants, are issued a financial identifier, called a "Social Security Number" (SSN), which functions as a corporate entity with its own potential to have liens placed against it. Of course, your SSN is not you, however, it legally *is* you for the purposes of their shell games. It is very difficult, if not impossible, to prove that you did not sign for loans, credit cards, or other contracted agreements from a bank, and the banks typically will not cooperate in resolving such instances of banking fraud since they are involved in every single one! Such crimes have a very low, nearly zero, rate of prosecution because the same people running the scam own the judges and the law.

Banks use whatever information you give them to violate your personal and financial privacy. In compliance with KYC (Know Your Customer) and AML (Anti-money Laundering) rules and regulations, your friendly, local teller analyzes you and your activities to determine if you are "suspicious." Bankers collect and sell your private information to their financial intelligence firms, like Equifax, where it is stored unencrypted in a centralized database to be stolen and sold on the Black Market.[12] Your answers to the nosy questions banks ask their customers, may be used by bank employees to commit forgery against you. They, also, work with third-party identity thieves to open fraudulent lines of credit in your name. In short, banks simply cannot be trusted with the information, which they demand on the inverted basis that *you* cannot be trusted.

By contrast, with Bitcoin no one wants to "get to know you as a customer." No one is eyeing your every move for supposedly suspicious behaviors. You do not have to be identified, fingerprinted, microchipped, vetted, verified, or background checked to obtain an account. No one cares who you are, where you come from, what you believe, or how you look. Bitcoin eliminates identity theft and subsequent banking fraud because you do not need personal and financial identity information to use it.

Cryptography and Witchcraft

Cryptocurrencies are so named because they act like paper currency and are based on digital cryptography. The word, "cryptography," is derived from two Greek words, which mean "hidden writing." The blockchain is based on different types of encryption used to facilitate and protect Bitcoin accounts and transactions. This modern mathematical encryption, performed using very fast computers, is based on historical cryptography, which is a subject familiar to witches and occultists.

Cryptography is used to hide information in plain sight, to securely transmit private information through a public channel. Witches and occultists have long used encryption to hide important information. In witchcraft, some symbols have power of their own, which is lent to a working. The need of an individual witch or a coven to encode information for secure storage and retrieval was very great during the centuries of the witch trials and remains very important today. Many famous witches and occultists, including John Dee and Aleister Crowley, have been expert cryptographers and there are many famous examples of encrypted occult books.

One such book is the *Voynich Manuscript*, named after the book dealer into whose hands it fell in 1912. It contains a cipher that has not yet been solved, however, experts believe it is a fifteenth-century Italian pharmacopoeia, containing information about herbs, astrology, and medical procedures.

The *Book of Soyga*, alternatively entitled "*Aldaraia*," is a sixteenth-century Latin manuscript, which contains procedures for conjuration, incantations, astrological information, and the names and relations of angels and demons. Some of it was deciphered by Elizabeth I's astrologer and spy John Dee. The undeciphered portion contains magic squares and it "makes numerous references to what are presumably mediaeval magical treatises, works such as liber E, liber Os, liber dignus, liber Sipal, liber Muriob, and the like."[13]

One of the most important organizations in the modern history of witchcraft, *The Hermetic Order of the Golden Dawn,* obtained some of its rituals from an encrypted document, known as the *Cipher Manuscript.* The cipher used to encode it came from the *Polygraphiae et Universelle Ecriture Cabalistique* (*Polygraphy and Universal Cabalistic Writing*) by Johannes Trithemius, published in 1561, which includes the first known publication of the Theban Alphabet.[14]

Aleister Crowley encrypted his own magical diaries and many of his published works, most notably *The Book of the Law,* contain cryptography. According to a researcher, Franz Bardon used a substitution cipher to communicate angelic names to his friends.[15]

Some witches and occultists practiced encryption either because they worked for the state, like Crowley and Dee, who were government agents, or because they had things to hide from the state and other powerful enemies. Although the practice of cryptography itself sometimes drew suspicion, partly because of its association with spirit communication. For instance, Trithemius authored another important book on cryptography, *Steganographia*, which was suppressed and because of this book, he "was accused of dabbling in the black art and holding converse with demons" and "had a very narrow escape of being burnt."[16]

Cryptography makes secure communication through the insecure channel of the Internet possible. Bitcoin is possible at this time because of the increased sophistication of encryption and recent advancements in the speed of computers to calculate complex algorithms involving very large integers.

The Blockchain: In Encryption We Trust

The blockchain is a progression, or chain, of encoded blocks of information. There are two basic styles of encryption: Linear and block. Block encryption is more secure than linear encryption because it employs ever-varying ciphers. The blockchain is secured by sophisticated digital encryption, predicated on probabilities, specifically, the time, energy, and computing capacity required to break the code it is based on, which renders fraud and forgery practically impossible.

Blockchain miners, are incentivized to engage in honest behavior and are kept in check by a consensus of the participants. When a new transaction occurs, an entry is made on the blockchain, a virtual ledger, which generates and assigns an encrypted code, called a "hash," that compresses information using a hashing algorithm, thus creating a progressive, chronological chain of transactions, which are vetted for authenticity by all miners in the system.

Each Bitcoin account involves two keys. The term, "keys," is a metaphor for strings of letters and numbers, mathematically connected to each other, which are determined when the computer solves an algorithm that amounts to nothing more than a complex mathematical puzzle with very limited possible solutions. One key is public and the other private, known only by the owner of the account and must be kept secret by the owner. This private key decrypts the public one and is used to send funds from the account. The public key is used to receive funds.

You can freely create as many key sets, alternatively referred to as "Bitcoin accounts" or "Bitcoin addresses," as you need. No identity verification is needed to obtain a cryptocurrency account and either send or receive funds with it. The only proof of your ownership is your possession of the private key to your account.

Bitcoin is Pseudonymous

While your private key is secret and known only to you, there is no secrecy inherent in the original Bitcoin blockchain itself. Bitcoin is *not* anonymous, despite mainstream media reports to the contrary. Instead, Bitcoin transactions are conducted pseudonymously, and unless the user takes extra steps, they can easily be traced. The blockchain ledger is a permanent, public record of transaction amounts, the date and time of processing, and the account from which the funds originated. Therefore, it is up to the user to secure his or her own relative anonymity.

This too public characteristic of Bitcoin has been remedied in subsequent altcoins, in particular, Monero, which defies tracking and tracing by making it impossible to see any user's balance but your own. Overall, the Monero project (getmonero.org) best represents the philosophy and mission of Bitcoin because it is on the cutting edge of innovation for financial privacy.

Mainstream media deception about Bitcoin's traceability and lack of anonymity may very well be intentional and not the result of ignorance. There are corporations that track all users' activity on the Bitcoin and altcoin blockchains, and at other websites that facilitate the use of cryptocurrencies. They compile and sell this data to your enemies.

Therefore, always use a VPN (Virtual Privacy Network) to cloak your IP address when interacting with any cryptocurrencies.

You are in Control

With Bitcoin, you control your own accounts and are in charge of your own security. You own and take responsibility for the things that belong to you. For many people, this is a novel idea, almost too shocking to grasp.

Imagine if more individuals began taking control of every aspect of their own lives. How quickly the Holy Brotherhood's system would disintegrate! Sadly, independence really is not for everyone. It never has been. The masses love authority. But take heart because you do not need to expend precious energy trying to convince the whole world of the importance of Bitcoin. Fortunately, Bitcoin requires neither their cooperation nor their good opinion to work for those select few who understand its true purpose.

Formal Education About Bitcoin and the Blockchain

Several universities offer online courses on Bitcoin, the Blockchain, and Cryptography, including Berkeley, Stanford, and the University of California.[17] Coursera (Coursera.com) offers an excellent online course on Bitcoin from Princeton University, called "Bitcoin and Cryptocurrency Technologies," (coursera.org/learn/cryptocurrency), which helps you understand the mathematical theory that supports the encryption of blocks of code on the blockchain, how the private and public key pairs are calculated, how the blockchain defies forgery, and how to protect your privacy when you use it. Stanford University offers especially good courses on Cryptography, called "Cryptography I" and "Cryptography II," through Coursera, (coursera.org/learn/crypto) which are useful in understanding the blockchain and secure digital communications, in general. Audit these courses free of charge or enroll in a course to earn a certificate for a fee.

In addition, the Internet Archive (archive.org) hosts digital versions of rare books on the subject of historical cryptography, including primitive block ciphers and mathematical cryptography, such as the following:

Cryptography, André Langié, Transl. J.C.H. MacBeth, Cryptography, Constable & Company Limited, London, Bombay, Sydney, 1922.
https://archive.org/details/Cryptography_593

Cryptomenysis patefacta; or, The art of secret information disclosed without a key. Containing, plain and demonstrative rules, for deciphering all manner of secret writing. With exact methods, for resolving secret intimations by signs or gestures, or in speech. As also an inquiry into the secret ways of conveying written messages, and the several mysterious proposals for secret information, mentioned by Trithemius, &c.," Falconer, John, D. Brown, London, 1685.
https://archive.org/details/cryptomenysispat00falc

Buchmann, Johannes A., *Introduction to Cryptography,* Second Edition, Springer Science+Business Media, LLC.
https://archive.org/details/IntroductionToCryptography

Mollin, Richard A., *An Introduction to Cryptography,* Second Edition, Boca Raton: Chapman & Hall/CRC, 2007.
https://archive.org/details/An_Introduction_to_Cryptography_Second_Edition

Bitcoin and the Blockchain as a Paradigm Shift

Understanding Bitcoin induces a mental shift and a new way of looking at the world and thinking about common things, even beyond money, business, and economics. When the beauty of the blockchain is grasped for the first time, it opens up a world of new possibilities, previously unimagined. It is truly a liberation of the mind, which allows you to conceive of doing many things in a completely different way.

In the paradigm of the previous age, many people desirous of equity and justice directly confronted authority. They tried to "fight the power," and openly protested in an attempt to illicit even a small measure of compassion from their enemies. Many expended a lot of energy in this struggle, but accomplished nothing more than reshaping the bars of the prison into which they were born.

By contrast, Bitcoin's permissionless nature inspires circumvention rather than confrontation. It is a total abrogation of authority, disengaged from the power structure, existing outside of it, almost as if it exists in another dimension.

CHAPTER 8
WHAT ARE MONEY AND CURRENCY?

"What is money?" "What is currency?" These seemingly simple questions have complex answers because the meaning of the word, "money," has been intentionally altered to suit the purposes of the Holy Brotherhood and mislead outsiders. Long ago, the true meaning of the word, "money," and the difference between "money" and "currency" became esoteric knowledge.

There is power in a word. The black magicians of the Holy Brotherhood know this very well, which is why they like to manipulate the names of things and their meanings. Therefore, it is important to consider the words used when describing things, so that they are always described accurately. When wrong words are used repeatedly, its like a trickle of water wearing a groove in the soil that, over time, becomes a river of confusion. It can happen easily in a short time, which is why many people now say "tin" when they mean aluminum, or "cement" when they mean concrete. Similarly, people have learned to incorrectly call bank IOUs, such as Federal Reserve Notes, "money."

Money

Money is a unique unit of value that measures labor, energy, and production. Historically, it is the weight measure of something valuable, such as gold. Money possesses some intrinsic value of its own and is relatively rare.

In the earliest known history of money, the term refers to coinage, which was described as a unit of measure or weight. Originally, a pound, a peso, a shekel, and a dollar all described the weight of something of value. According to Blacks Law Dictionary, published in 1891, the dollar is "The unit employed in the United States in calculating money values. It is coined both in gold and silver, and is the value of one hundred cents."[1] Real money is both an intrinsically valuable measure of wealth, such as grain, cattle, gold, or land, and a medium by which to store and transfer it.

Currency

By contrast, currency has no intrinsic value, although it is an instrument of value transfer. Federal Reserve Notes are best described as "debt-based currency" or "usury-based currency" because they circulate and increase debt, instead of circulating and increasing wealth. The act of replacing actual money with usurious IOUs is another example of the Holy Brotherhood stealing a genuine thing and replacing it with a fraudulent one. By means of this language manipulation, teaching the public to call usurious, debt-based currency "money," the Holy Brotherhood deceives nations and individuals in order to siphon off their real wealth.

Ancient Terms Related to Money

Everyone familiar with the Old German runes, the "Elder Futhark," commonly used in modern rune casting, knows that "fehu," the first rune of the Futhark, means "cattle" and is regarded as the rune of wealth. Cattle are genuine wealth, which have value both as food, as livestock, and for trading or bartering. The more modern English words "fief" and "fee" are derived from "fehu."

The Latin word for money is "pecunia," which is derived from the Latin word, "pecas," which means "cattle." Also, "pecunia" means "fund," "wealth," "property," and "bribe." In English, the word "pecuniary" means "monetary."

The Roman goddess Juno Moneta was the protector of money coined at the mint inside her temple at Capitoline Hill in Rome. The modern English words "money," "monetary," "monetize," and "mint" come from her name. In modern Italian, "Moneta" means "coin."

The Bankers' Linguistic Shell Game

A bank note is an IOU, a promise from a bank to pay real money, something that has intrinsic value, to the bearer of the note. At one time, bank notes, such as the one depicted in the illustration below, which bears an image of Juno Moneta guarding a treasure, could be redeemed at a bank for gold.

This $5 bank note from 1861 Confederate States of America depicts Moneta holding a rod with two serpents entwined around it, guarding a treasure chest while casting backward glance at a merchant ship. A sailor is depicted to the left. Image Source: Smithsonian National Museum of American History.

People began to erroneously refer to bank notes as "money" and continued to do so even after they became empty promises, not worth the paper they are printed on. It is not possible to take a Federal Reserve Note to a bank and exchange it for anything of real value. Even the coinage has been nearly devoid of intrinsic value since 1964, which is when the production of silver-containing coins ended.

Even at the time of the gold standard, Federal Reserve Notes never really had any guaranteed value because of the central bankers' criminal practice of fractional reserve lending. Furthermore, backing fiat, debt-based currency with gold only permits the central bankers to defraud the public more quickly and effectively. History shows that, when a gold standard is implemented, the Holy Brotherhood steal citizens' gold by color of law to back the currency, shortly after which all the gold disappears into the night.

Central bank notes are referred to as fiat currency. Fiat currency is that which exists because of a decree, mandate, or authorization by an entity that presumes authority. Biblically, fiat is God's act of creation in the words he spoke, "Let it be done" and it was done. "Fiat" is Latin for "to let," "to make," or "to do." For example, "Fiat lux," is Latin for "Let there be light." Fiat currency, having no intrinsic value of its own, is only buying power backed by an authoritarian decree and faith in the one who made it, whose power historically comes from God.

Cryptocurrencies

Cryptocurrencies are pseudo-currencies, which transfer value, however, they do not fit the definition of either money or currency. Cryptocurrencies are not money because they have no intrinsic value. They are not fiat nor are they legitimate currency in that they are not spoken into existence by an authority figure. Cryptocurrencies are brought into being, not by the speaking of a holy word or decree, but by the open source digital code of the blockchain.

The Legal Definition of Bitcoin

Attempts to legally define Bitcoin have produced a variety of confusing results. Some say it is money; some say it is a commodity. Certainly, it functions like currency, although it is not truly currency because it is not issued by a government or a banking authority. A few Federal judges have ruled that it is not money and, therefore, cannot legally be treated as money in court proceedings involving such matters as money laundering.

Bitcoin and altcoins are called "property" and "commodities" by the I.R.S. Although, Bitcoin's intangible nature defies the definition of both since it is never in the user's possession, rather it is a string of code in cyberspace.

When black magicians cannot correctly name something, they cannot control it. Thus, the Holy Brotherhood's inability to define Bitcoin fortifies it.

CHAPTER 9
WHY FINANCIAL PRIVACY IS IMPORTANT FOR WITCHES AND OCCULTISTS

"Whenever we read the obscene stories, the voluptuous debaucheries, the cruel and tortuous executions, the unrelenting vindictiveness with which more than half the Bible is filled, it would be more consistent that we call it the word of a demon than the word of God. It is a history of wickedness that has served to corrupt and brutalize mankind."

- Thomas Paine, *The Age of Reason*, 1794-1807.

Financial domination is the heart of the Holy Brotherhood's power. Money means the ability to live in the world, which is why they financially constrict their enemies. It is a highly effective way to weaken them, destroy their essence, subjugate, and control them. While there may be some protection in forming witchcraft-based religions, it does not protect the very existence of witches and occultists against their secular persecution, which is most often financial in nature.

Economic abuse is devastating, but much more subtle than public immolation, so it typically goes unchallenged even when it is blatantly illegal. For instance, despite a Federal law prohibiting discrimination, the Holy Brotherhood overtly restricted women from obtaining banking services, especially lines of credit, until the 1970s. Presently, witches, occultists, and often pagans, especially any who are out of the closet due to the necessity of conducting business, are overtly denied the banking services necessary to do so and denied equal protection under the law.

In the modern Western world, there are few overt anti-witchcraft laws. Most of the time, we experience covert legal restrictions on the traditional practices of witches and occultists, written without an express prohibition on witchcraft itself. The effects of this are nearly the same as having anti-witchcraft laws.

This is the case in the United States, where witchcraft is primarily protected under the First Amendment, yet there are many restrictions on practices inherent to traditional witchcraft at both the Federal and local levels. It is similar in England, where once the Witchcraft Acts were repealed, they were replaced with laws restricting the practices of witches. Similar laws restricting witchcraft, not in name, but in practice, exist throughout the British Commonwealth.

For instance, it is a well-known fact among the common public that healing and medicine are important aspects of traditional witchcraft because this was frequently given as the cause for our arrest during the years of the witchcraft trials. Even when there are no explicit laws against witchcraft, witches are still restricted from the practice of healing, preparing medicines, selling healing agents, and even administering them free of charge. They have progressively enacted laws against the gathering of some herbs or the purchase of them, whether for personal use or in making medicines for others. Moreover, as a result of centuries of their informational and institutional controls, most authoritarians, whether religious or secular are propagandized to believe that this is good and desirable.

Most assuredly, the effort to convince witches and occultists that we have no right to profit by our abilities is psychological warfare perpetrated by our historical enemies. It has the bloody fingerprints of the Holy Brotherhood and their depraved ideas about ethics when it comes to women, witches, and money all over it. It is highly unethical, greedy, oppressive, and exploitative to expect someone to work, to use their time, energy, skills, and knowledge base, without just remuneration. It is enslavement, which is the Holy Brotherhood's legacy. To this day, many witches have limited economic options, yet are restricted from performing our own esoteric arts and sciences altogether or are expected to do so free of charge. In public hearings of this decade, when attempts have been made to remove statutes against "fortunetelling," "gypsies," and "magic arts," witches are taunted about needing fair compensation for services rendered.[1]

In some cities and counties, witches and occultists are refused commercial licenses and may be arrested for conducting witchcraft-related business. Sometimes they are granted licenses at prohibitively high fees, in some instances $400 to $1,000, to obtain a license to read tarot cards. The licensing process may include an intrusive background check and being fingerprinted like a common criminal. Such licensing creates a permanent public record that identifies a witch or occult practitioner, which may result in employment discrimination, for which there is rarely sufficient, if any, legal recourse. Many witches and occultists must maintain secrecy for professional reasons, which is a deterrent against obtaining such licenses or engaging in public hearings or protests against restrictive laws.

Furthermore, witches and occultists are subject to regulation on the basis of legal requirements surrounding a legal definition of religion that is modeled on the organizational structure of the Holy Brotherhood's synagogues, churches, and mosques, such as having buildings in which to worship.[2] In this way, witches and occultists are denied legal protections extended to members of the Sinister Trinity.

Even in the seemingly free-enterprise world of online commerce, witches and occultists are denied the opportunity to do business or must do so with the threat of being shut down hanging overhead like the sword of Damocles. Traditional financial institutions may, at any time, close their customers' accounts and confiscate their funds with no recourse for the wronged party.

Presently, a very few interrelated corporations monopolize banking and online payment processing, and only one monopolized online payment processor dominates every single online shopping cart system. As an exercise of their absolute power, in 2012, on a whim and with very short notice, they suddenly forbade the sale of a particular genre of books and threatened the sellers with the deactivation of their accounts, which caused great distress and loss for book distributors, writers and publishers.[3] Consequently, the livelihoods of innocent people were threatened, many of whom were not even in the targeted group. Only after months of a public outcry, the payment processor relented, but the damage was done.[4]

Similarly, there have been attempts by major online auction websites and other third-party corporations to restrict the sale of witchcraft-related items or services. For example, Etsy began restricting the sale of metaphysical items at their site and reportedly shut down metaphysical stores without warning in 2015.[5] Reportedly, similar restrictions have been in place at Ebay since August 2012.[6]

Under one pretext or another, some payment processors explicitly prohibit the sale of "occult materials," which includes typical metaphysical items, such as amulets and tarot cards.[7] These third-parties have the power to capriciously shut down businesses, terminating their accounts without warning. Additionally, they use vague terms in their "Terms of Service," which are purposely broad so that they may forbid anything at all, even rocks, books, or clothing they disapprove of.

Banks and centralized payment processors, also, shut down donations to controversial individuals and causes. Moreover, if you make a donation through them, your name and personal information is indefinitely stored in a database, which means that you could be publicly connected to that individual or cause, if not now, then years from now.

Witches and Occultists Targeted for Economic Abuse

It is usually the case that you do not know much at all about the banks or other corporations you deal with, but they certainly know a lot about you, and are, in fact, gathering data about you all the time.

When you go to your local bank and make a deposit, the teller can see how much money you have in your account. Bank employees have access to any other information you have provided including addresses, phone numbers, the names of other members of your family, your occupation, and your transactions.

Your financial transactions reveal a lot about you, including your beliefs, lack of beliefs, where you have been, where you go, the identities of your associates, and any political causes you support. It is important for witches and anyone else who might be subject to discrimination, harassment, or other crimes to maintain privacy. This is much easier to do when your name, address, and banking information are not associated with the things you buy or sell.

If you are lucky, you live in a place where there is tolerance for witches and occultists and you practice some form of witchcraft regarded as benign, but not everyone is so fortunate. In many areas, there is hostility toward anyone who is not a member of the dominant religion. Moreover, many traditional witchcraft and occult philosophies and practices, such as Satanism and black magic, are controversial even within the broader community of witches and pagans. If your form of witchcraft is deemed unacceptable by either of these groups, then you need even more personal and financial privacy. Even in countries like the United States, the United Kingdom, and Canada, in which civil rights are supposedly held in high regard, witches and occultists are forced out of their businesses, jobs, and homes, are threatened and are targets for vandalism and violence.

In the United States, publicized examples of this include the Seekers Temple (seekerstemple.com), presently of Florence, South Carolina, whose plight reportedly began in 2014 when they tried to go through the proper channels in Beebe, Arkansas, to obtain a license for their nonprofit organization and were stonewalled by the local government, harassed by a neighboring church and the police, threatened with imprisonment and murder, and denied equal protection under the law.[8] As is typical, the local city council appears to have used zoning laws as a hammer against them.[9] According to an account posted at their website, financial damage is part of the targeting, which continues even though they fled Beebe for their "personal safety" in 2016.[10] This pagan church's pleas for justice have been to no avail.[11]

On November 1, 2015, even before their official opening and after a disruptive Christian protest,[12] the front window of the Greater Church of Lucifer in the Houston, Texas suburb of Spring, was smashed and a large limb was cut from a tree, thus smashing its roof.[13] After enduring months of harassment and protests, on December 22, 2016, in an apparent attempt to appease their persecutors, the organization announced that it changed its name to "Assembly of Light Bearers."[14] On April 23, 2017, it was reported that the Luciferian church lost its lease after a final act of vandalism was committed at its headquarters and the landlord received death threats.[15]

In May 2017, it was reported that an English occult shop located near Gloucester Cathedral, called "Spellbound Witch School," was targeted, branded as "terrorists" and "devil worshipers," and its proprietors were receiving death threats, amid a string of incidents of harassment and abuse.[16] Despite the business having suffered criminal damages, they were unable to get cooperation from local law enforcement. The following month, it was reported that the police finally got involved when it "escalated into threats to 'burn the place down' – with the staff still inside."[17] The police released a CCTV photo of a person of interest relating to criminal activities, which occurred after a group had performed an exorcism outside the store.[18] As a result of this ongoing persecution, the owner was forced to install costly surveillance cameras, their customers have been subjected to harassment and intimidation, and they have had to remove the "store's social media" page.[19]

In Manitoba, Canada, a pagan store, called "Elemental Book & Curiosity Shop, Inc.," has reportedly been targeted by vandals on the basis that the owner is a witch who sells occult materials.[20] Reportedly, the police are reluctant to become involved because they consider the years of harassment and vandalism the owner has endured as minor "mischief" and "unimportant."[21] Interestingly, there is debate over whether Canada's "hate crime" laws would apply to the criminals since witchcraft does not fit the legal definition of "religion," which demonstrates the ineffectiveness of witches trying to secure inherent rights on the grounds of religion.[22]

What Seems Safe Today Might Not Be Tomorrow

The social environment is in flux, which means that what seems safe today might not be so tomorrow. For instance, 35 years ago, the owners of a bookstore in the Denver, Colorado suburb of Englewood, called "Isis Books & Gifts," could not have predicted that they would become targeted for vandalism and harassment because of their store's name. "ISIS" is the common name of a terrorist organization, which began receiving news coverage in 2014.[23] After enduring harassment and vandalism to their sign, they announced on December 29, 2015, that they changed their store's sign to "Goddess of 10,000 Names," in an effort to assuage threats and vandalism.[24]

The Witch Hunts are Never Far Behind Us

There are wide-eyed idealists who envision a world of progress in terms of human evolution, but this is not reality. The Western world is experiencing a resurgence of religious fanaticism in which women, regardless of stated belief and public prayer-making, are, once again, taunted as witches amid cries of "Burn her!" resurrecting fears of a return of "The Burning Times" among some pagans and women's advocacy groups.[25]

Reminiscent of the 14th century European witchcraft persecutions, such accusations as pagans promoting homosexuality to children, tarot readers harming children, and witches opening up demonic portals have resurfaced in ordinary places in the United States, as demonstrated at the Town Council Meeting held in Front Royal, Virginia, in 2014 to repeal a law against tarot card reading.[26]

In many U.S. cities and counties, law enforcement agencies have begun placing "In God We Trust" bumper stickers on squad cars and quoting *Romans 13* at their official websites, reminding us all that their authority to rule comes from God and that police are his avenging ministers against heretics. If you live in one of these areas, you have little reason to believe that you will receive cooperation from them should you become the victim of a crime.

Internationally, the persecution of witches is a growing problem. The U.N. estimates the number of murders of accused witches to be thousands per year worldwide with a concentration of activity in the Democratic Republic of the Congo, India, Nepal, South Africa, Ghana, Tanzania, Angola, Papua New Guinea, and Nigeria.[27] The targets are most often women, often elderly ones, and children.[28] Accused witches are not always murdered, burned alive or beaten to death, some are driven out, and some are imprisoned for decades in camps.[29] Witchcraft is punishable by death in countries dominated by Islam, such as Saudi Arabia, in which women are reported by neighbors or employers, investigated, arrested, and convicted in secular courts for the crime of practicing witchcraft.[30] Contemporary accounts of religious fanatics persecuting witches read nearly identically to those of past centuries and they are becoming more common in English-speaking nations, especially those with large communities of immigrants from countries in which witchcraft is illegal.

Never underestimate the Holy Brotherhood's penchant for cruelty when they perceive that someone is not properly venerating their fictional God or his secular authority. Our historical enemies have survived the centuries with us, only their titles and methods of persecution have changed a little bit. They really want to commit their past atrocities openly and without restraint, and they will do it, again, as soon as they think they can get away with it.

Witches and the "Social Contract"

In an attempt to legitimize the secular power of government authority, previously said to come directly from God, European philosophers of the seventeenth to the nineteenth centuries developed the concept of a "social contract" or "social agreement" According to this philosophy, because you are born in a place, you have consented to forfeit some of your inherent rights and freedoms in exchange for the alleged protections of the state and whatever rights it decides to allow you. This secular form of asserting the state's authority over you assumes that the system works most of the time. While such systems may work for the men who created them, they have only worked to the destruction of witches and occultists. Just as there is no God, there is no such secular obligation, especially for witches.

Still, among modern witches, there are those who believe it is their civic duty to work within the system. Some try to change laws to make them more fair, tolerant, and inclusive, in an attempt to peacefully coexist with our mortal enemies. They petition, give interviews to local media, protest, and sometimes bring lawsuits in an attempt to force changes. Apart from being costly, time-consuming, and dangerous, this is often a fruitless course of action. Moreover, doing this only acknowledges and gives credence to a wholly illegitimate authority. If witches could achieve the tolerance of our enemies by such methods, the Holy Brotherhood and the Men of God would have taken their boots off our necks a long time ago.

Furthermore, attempts to gain legal protection on religious grounds are insufficient because being a witch or occultist is not equivalent to being a member of a religion. Primarily, the Holy Brotherhood's biggest concerns about witches have always revolved around what witches are and can do, rather than what we believe. Our nature as religious heretics is only a secondary reason for their ongoing persecution.

Witches and Submission to Authority

As illustrated by the above examples of contemporary persecution, witches and occultists often get into the most difficulty when they try to follow the Holy Brotherhood's rules, which are designed to protect themselves. Our enemies believe that all authority comes from God. They see themselves as ministers of their God. Therefore, acknowledging their fraudulent authority by trying to comply with their rules feeds their delusions of power over you. It convinces them all the more that they have the right to subjugate you.

Increasingly, attempts to peacefully co-exist or promote religious tolerance fail. Such efforts were better received 50 years ago than they are now. At this time, a high level of authoritarianism prevails because the old institutions are in peril due to changing cosmic frequencies. Their God has one foot in the grave. Witchcraft is the future. As a witch, occultist, or pagan, you represent the Apocalyptic future they fear most.

Never underestimate your historical enemies. According to their doctrines, you are a soulless subhuman, a living devil or a vile temptress, and the embodiment of Satan. Appealing to their sense of justice, fairness, or humanity does not work because they have none where you are concerned. Christian talk of love and charity does not apply to witches, it only applies to themselves. Likewise, it is futile to appeal to law enforcement, God's avenging ministers against sinners, when you are targeted by criminals. From their perspective, the criminals are doing God's work, and you deserve death by immolation or hanging.

The compassion of the Godly lies in the belief that they are saving your immortal soul when they murder you. Since openly killing people on the basis of witchcraft is, more often than not, frowned upon these days, they use the power of their long-standing monopolies to harm witches, denying you full economic participation by withholding their services.

But, the Holy Brotherhood cannot economically abuse you when you use Bitcoin and altcoins peer-to-peer. Bitcoins enable you to establish your brand without outing your identity and to protect the privacy of your clients and customers. You protect your privacy when you make purchases with it. Dealing directly, one-to-one, with buyers and sellers using Bitcoin negates the possibility of threats and discrimination by banks and other third parties. It can help you lower your public profile and reduce the possibility of being targeted for discrimination or violence, now or in the future.

Whether the Holy Brotherhood claims a religious or secular right to rule you, God is their original and ultimate fraud, "the Alpha and the Omega" of their great swindle. You immediately remove yourself from it when you operate outside their "In God We Trust" system. You claim your birthright when you do not seek the permission of presumed authorities.

Bitcoin is the antidote to the Holy Brotherhood's venom because it is designed to let you do just that. It restores your economic self-determination and inherent right to privacy. To use it well, you must come to *know* it, *dare* to use it, and, as much as possible, *keep private matters private.*

CHAPTER 10
WITCHES, TECHNOLOGY, AND
CRYPTOCURRENCY ADOPTION

Thanks to the marketing efforts of a major New Age publishing house, whose name is known by every witch and occultist in the English-speaking world, witchcraft is widely regarded as an earth-centered spiritual practice focused on nature. This is at best, of course. While it is true that witchcraft concerns itself with matters pertaining to the natural world, this is by no means the whole picture and is a simplified, symbolic explanation for more complex and important things. One could spend a lifetime studying and experimenting with witchcraft and never fully plumb the depths of the subject.

Fire, air, water, earth, ether, and the magnetic union of the divine masculine with the divine feminine to create life are basic, esoteric concepts in occultism. The names of the elements are not meant to be taken literally, nor worshiped in personified forms as gods and goddesses, nor to be merely celebrated in ritual, rather they represent natural forces, which can be manipulated to produce a power, an energy, such as electricity or plasma. Essentially, witchcraft is the application of an occult science, which may be used to solve problems in the physical world by manipulating forces that seem, at least, to most people, to be non-physical.

Not only is there nothing new under the sun, there is nothing but old technology, old know-how, stolen from our ancestors by the Holy Brotherhood and hoarded for millennia, so the common people would remain ignorant, superstitious, and manageable. This thousands-of-years-old technology has been rapidly unveiled since the mid-nineteenth century, and is passed off as "new technology" or a "discovery," when, in fact, it is very, very old. Simultaneously, the equally ancient science behind it and the esoteric sources of this information are suppressed by the Holy Brotherhood's subsystem of education. The schools and universities that most people believe teach science actually do everything they can to obfuscate it.

Occultism is the Basis for Modern Communications Technology

Modern communications technology is rooted in spirit communication and was pioneered by men and women with a profound interest in the occult. Communication technology involves the transmission and reception of messages and images across long distances, using frequency harmonics and objects of influence, which are concepts commonly used in the practice of witchcraft. These concepts and their underlying esoteric science are detailed in this author's book, *How to Write Your Own Spells and Make Them Work*, a workbook of spell crafting.

Spirit boards and automatic writing planchettes were the earliest technology for long-distance communication with both the dead and the living. They were superseded by telegraphs, telephones, radios, and television. In the 1920s, Thomas Edison worked on an apparatus for communicating with "deceased personalities."[1] Information about his work on this invention, called the "Spirit Phone," has been suppressed and excised from his official biography.

Nikola Tesla, the true inventor of the radio, the Tesla coil, and Alternating Current electricity, whose work laid the foundation for computers, wireless technology, and radar systems, was familiar with Vedic philosophy and used Sanskrit terminology, familiar to occultists and students of Theosophy, to describe the science behind his inventions. Sanskrit is the language of the ancient text, the *Mahābhārata* ("Great Epic of the Bharata Dynasty"), a sacred history of India, which describes ancient technology, including the construction and use of atomic weapons and flying machines, and appears to be the true source of many modern scientific "inventions" and "discoveries." Much of Tesla's work is suppressed.

Early twentieth-century electronic devices, such as television sets and radios, relied on vacuum tubes, which were pioneered by the physicist and spiritualist, Sir William Crookes, who, also, scientifically experimented with psychic forces and spirit communication, as described in his book, *Researches in the Phenomena of Spiritualism*, published in

1874.[2] He designed the residual ectometron and ectoplasmic batteries, which aide mediums in generating enough ectoplasm to produce a physical manifestation of a deceased person.[3]

Electromagnetic wave forms and radio frequencies are the essence of homeopathy and radionics. Similar harmonic resonance and energy field manipulation technology forms the basis of other energy healing methods, such as those used by Franz Anton Mesmer, in which the practitioner exerts an electromagnetic influence on the body's bioplasmic fields, whether through the healers' hands or mental emanations.

Contemporary occult researchers use common cameras, audio recorders, and video recorders to verify the experiences of psychics and mediums. Electromagnetic Frequency (EMF) detectors and heat-detecting cameras may be used to verify the emanations that those with highly developed sensory perception can detect that other people cannot. Radio equipment, such as the Spirit Box, is used as a medium of spirit communication by some paranormal investigators.

In short, there is every reason why witches and occultists might be especially comfortable with modern communications technology and electronics.

Witches and Computer Technology

In the 2006 edition of her book, *Drawing Down the Moon: Witches, Druids, Goddess Worshipers and Other Pagans in America,* Margot Adler references a survey of "Pagans and Wiccans" she conducted in 1985, which indicated that the largest category of their employment (16%) involved computer technology or software development.[4] While this may surprise those who believe witchcraft is another form of tree hugging, it probably does not surprise most witches.

Not only does occultism provide the basis for modern communications technology, but many new members find their way to the witchcraft community by means of this technology. For decades, witches have made great use of the Internet for communicating with each other. It allows us to share experiences and learn from each other in ways that were simply not possible before.

Moreover, witches have a lot in common with other early adopters of new technology, who are often regarded as unconventional. For instance, nineteenth-century photography enthusiasts were regarded as strange. So were early automobile owners. Similarly, people interested in computer technology from the 1970s to the early 1990s were regarded as odd and women, in particular, were belittled and nagged by Luddites who thought girls should find better things to do with their time. Similarly, Bitcoin users are sometimes ridiculed and derided as technology-obsessed oddballs and annoyances.

Such disparagement is especially unlikely to deter witches and occultists, who are motivated to seek alternatives to a system that denies us full participation and exposes us to dangers. Witches, as members of a historically persecuted group, have a greater than average need for anonymous means to communicate and conduct business. Cryptocurrencies present a new avenue for you to transact business at a distance, in greater privacy, without the risks inherent in traditional banking. It is no wonder that a great number of witches, occultists, pagans, and other heretics now offer their goods and services on the Internet in exchange for Bitcoin and altcoins.

Adoption of Bitcoin

Many merchants accept Bitcoin because they simply want to offer their customers as many different payment options as possible. It is popular with people who are concerned about the alarming degree of warrantless, unconstitutional, mass surveillance that is conducted by means of the banks and who understand the role of encryption in restoring privacy. It is loved by civil liberties advocates, especially those comfortable with information technology, however, Bitcoin is most quickly adopted by those denied banking services or who lose their trust in the banks.

Between 2014 and 2015, U.S. credit card companies canceled the merchant accounts of Mexican and overseas pharmacies, which many Americans use to obtain pharmaceuticals at reasonable prices and without perverse intrusions on personal privacy.[5] Many of these pharmacies are among the early adopters of Bitcoin and altcoins. Consequently, their prices dropped dramatically because they are no longer passing on large banking fees to their customers.

About the same time, members of the legalized marijuana industry in the United States developed their own altcoin, called "Potcoin" (potcoin.com) because most banks refused to work with them. According to its official website, Potcoin is "ultra-secure" and allows members of this agricultural community to "interact, transact, communicate and grow together."[6] Members and supporters of this industry were, also, among the first to adopt Bitcoin and Monero.

Companies that have been threatened and, consequently, lose trust in third-party payment processors are, also, quick to adopt Bitcoin. In 2016, Seafile, a German-based, fully-encrypted cloud storage company similar to DropBox, was accused of violating PayPal's terms of service for failure to monitor and report on the contents of their customers' files and personal data, although doing so violates German and European privacy laws.[7] At the demand of their customers, Seafile began accepting Bitcoin.[8] PayPal later apologized for their "mistake" and offered to reinstate Seafile's account, but they declined the offer, stating that the incident "has shattered our trust in PayPal."[9] Instead, they continued accepting Bitcoin payments by means of a third-party, BitPay,[10] which partners with PayPal to accept Bitcoin as payment.[11]

Categories of merchants who have experienced problems with traditional payment processors or who run the risk of doing so often accept Bitcoin as payment. This includes psychics, tarot readers, spell casters, astrologers, spiritual counselors, healers, root doctors, and those who offer online classes, coaching, occult merchandise, alternative medicine, nutritional supplements, and self-defense items.

According to an article, "Who Accepts Bitcoins as Payment?" at 99Coins.com, a long list of major companies accept Bitcoin as online payment, including, Wordpress, TigerDirect, NameCheap, BigFishGames, Bloomberg.com, Whole Foods, Dish Network, and Overstock.com.[12] Lots of smaller companies, such as Go Green Solar (gogreensolar.com), which sells solar panels; and Smyth/Cid Water Filtration Crocks (waterfiltercrock.com), which sells beautiful, handmade stoneware water filtration crocks, accept Bitcoin. Numerous web hosting and domain providers accept Bitcoin and altcoins, including Glow Host (glowhost.com), Hawk Host (hawkhost.com), Host Winds (hostwinds.com), and Name Cheap (namecheap.com). Some online gaming companies, VPN services, email services, Cloud services, coin and collectible shops, gold bullion dealers, freelancers and increasing numbers of businesses of all kinds accept Bitcoin and some alt-coins as payment. Conduct a web search for *"Buy a [type of product or service] with Bitcoin"* to fetch results for companies that accept it in exchange for particular products or services.

Around the world, brick-and-mortar stores are offering Bitcoin as a person-to-person payment option. In 2013, it was reported in the UK Guardian that a number of brick-and-mortar stores in the Kreuzberg district of Berlin began using Bitcoin.[13] A major European retailer, Alza, now accepts Bitcoin and has Bitcoin ATMs installed in some of its stores in Prague.[14]

Since April 1, 2017, when the Japanese government officially recognized Bitcoin as a form of payment, its acceptance increased. Reportedly, tens of thousands of Japanese businesses already accept Bitcoin and many more are expected to get on board in the near future.[15]

Bitcoin is integrated into ecommerce shopping cart systems,

commonly funneled through a third party that partners with PayPal or a centralized exchange. In particular, PayPal partners with other third-parties, including centralized, custodial cryptocurrency exchanges, to allow merchants to accept Bitcoin payments. The benefit of using a third party is convenience. But, the drawback is that it negates the privacy and security benefits of Bitcoin's peer-to-peer decentralization, creates financial and privacy risks, thus reducing Bitcoin to just another way to accept and make payments.

Bitcoin and altcoins are commonly accepted as donations. It is very simple to place your Bitcoin or altcoin account addresses and QR codes on your website, blog, social media account, or anywhere else, and immediately begin accepting payments. This was most famously done by WikiLeaks after PayPal and Visa "permanently suspended" their account on December 3, 2010, allegedly acting in accordance with agencies of the U.S. government, in an act of "digital McCarthyism."[16] The move came days before it was revealed the WikiLeaks was preparing to "expose tens of thousands of files relating to abusive practices in US financial institutions."[17]

Presently, the number of people who use Bitcoin may be in the millions, while the number of those who use altcoins may only be in the thousands, however, Bitcoin and altcoins do not have to be widely adopted to be beneficial to a community of users.

Using Bitcoin and Altcoins Protects and Enriches the Witchcraft Community

The information age has made it very easy for witches and occultists to communicate with one another anonymously, which has been very beneficial, leading to an explosion of interest in all aspects of the occult, and reducing the risks involved in meeting in person. As a result, a large community of witches and occultists of all kinds exists on the Internet.

Bitcoin is the Internet's own currency, designed especially for web transactions. Using cryptocurrencies helps witches and occultists escape many of the dangers of public exposure and censorship. Bitcoin and altcoins help buyers and sellers of occult products and services maintain their privacy and reduce the risk of being targeted for discrimination and harassment.

Using Bitcoin means that vulnerable individuals and businesses are no longer at the mercy of potentially hostile banks and other corporations. It removes the possibility of intrusion by the Holy Controllers, who usually have some fabricated excuse for sticking their noses into your personal business or denying you something you need, whether it is a license, a permit, an account, or equal protection under the law.

When you buy or sell peer-to-peer with Bitcoin or altcoins, you do not have to expose your private or financial information to any third party. You only provide the minimum information required to complete the transaction. When Bitcoin is used correctly it is difficult to track your activities and connect them to your identity, thus preserving you from being inadvertently outed from the broom closet. Your privacy and security are even more assured if you use a privacy coin, such as Monero. Using cryptocurrencies with discretion helps protect your inherent and legal right to personal privacy. When you protect your personal privacy, you protect your other inherent and legal rights.

CHAPTER 11
THE PARADIGM SHIFT

"In order to DARE we must KNOW; in order to WILL, we must DARE; we must WILL to possess empire, and to reign we must BE SILENT."

- Eliphas Levi, *Transcendental Magic, Its Doctrine and Ritual*, 1854.

While Bitcoin is a chance at liberty for the wise, it may very well be a trap for the foolish. It has already proven troublesome for some who failed to grasp its underlying philosophy, often because they think it is unimportant or just claptrap. This is their folly and where they fail, you will succeed all the more.

Overall, the biggest potential hazard for users of Bitcoin and altcoins is the failure to make the necessary paradigm shift, to remain stuck in old ways of thinking and doing things. The purpose of Bitcoin is to liberate the oppressed from the clutches of the Holy Brotherhood by means of encryption. When you maintain your right to privacy, your other liberties are preserved. To use Bitcoin and altcoins for privacy, please, consider the following recommendations:

Keep your private keys private. When you acquire an account, there will be a private key. Whoever possesses it owns the account. Your private key is the only proof of ownership you have. *Keep your private keys private.* This is the single, most important rule in Bitcoin and altcoins and if you always adhere to it, you will keep your accounts safe. Nearly everything else that follows in this chapter is merely an extension of this one cardinal rule.

Do not trust a trusted third party. Do not combine decentralized, peer-to-peer cryptocurrency with centralized, trusted third parties. By doing so, you lose the benefits of Bitcoin while incurring all the risks of banking and then some, since you do not even have the scant promise of government regulations to protect you. While it is often more convenient, anytime you give anything of value to someone else, you risk losing it.

Retain the financial autonomy Bitcoin provides you and do not, whether willfully or by deception, allow it to fall into the hands of one of Gods' trusted third parties. For maximum security and privacy, use Bitcoin peer-to-peer, as intended.

Never mix Bitcoin with banking. Banks and payment processing companies are trusted third parties, which are especially hostile toward Bitcoin and altcoins. Traditional financial institutions have confiscated the funds and closed the accounts of their customers who buy or sell cryptocurrencies. Mixing banking with Bitcoin is risky and negates its privacy and security.

Beware of banker-involved blockchains. Banks have become involved in blockchain-based cryptocurrencies in an effort to regulate and "fix" the things about it they think are wrong, for instance, the lack of authority and "trusted third-party" centralization. Bankers are involved with two main altcoins: Ripple and Ethereum.[1] Ripple uses a centralized system of transaction in place of mining and Ethereum is notorious for spawning ICO (Initial Coin Offering) scams.

Adulterated versions of the original blockchain can be compromised, manipulated, and brought under the Holy Brotherhood's control while giving the false appearance of carrying out the philosophical mission of Bitcoin. Whenever you see the Holy Brotherhood funding a project, when it contains any aspect of their signature architecture of pyramids (authority and hierarchy) and circles (centralization), or implements any of the traditional, fraudulent schemes from their ancient playbook, this should raise a big, red flag. *Approach with caution.*

Do not believe the delusional assertion that Bitcoin is a commodity. It is dangerous to both your finances and your privacy to believe that cryptocurrencies are commodities. The Holy Brotherhood's designation of cryptocurrencies as "commodities" is a deception that facilitates pump-and-dump schemes, ICO scams, High-yield Investment Programs (HYIP) and other Ponzi schemes, racketeering, and other forms of organized criminal activity. Many technologically illiterate and financially unsophisticated investors have lost large sums of money on such scams already.[2]

The most common type of cryptocurrency-based scam is the ICO, in which the scammers sell their digital tokens for U.S dollars or other local currency. ICOs are phony investment scams that resemble IPOs (Initial Public Offerings), which are stocks issued by private companies or corporations to raise investment capital. Typically, using a built-in feature of the Ethereum blockchain, a scammer creates a cryptocurrency "token" with a snazzy name, which they purport serves some questionable function. They set up a website for their fictitious startup company, write a white paper, create videos portraying themselves as hard-at-work computer geniuses, and give interviews to highly paid promoters. If the ICO is a popular one, the token may briefly be listed at popular custodial cryptocurrency exchanges, an event that will be loudly celebrated by the promoters.

Commonly, the promoters are minor celebrities and likable online personalities on social media websites, who often fail to disclose their relationship to the phony CEOs. They portray themselves as insiders-in-the-know and may employ an array of charlatans, including phony investment experts, fraudulent psychics, and prognosticating bots to encourage naive speculators to buy the ICO token, quickly pumping up its exchange rate. The ICO scam is presented as "insider information" and a "once-in-a-lifetime opportunity" that only a few, lucky people will get news of at the right time to make a fortune. The promised return on investment is often a large percentage within a day or two. Once they pump up the value of their token and collect a large sum from their "investors," the group's real insiders sell off, or dump, their cryptocurrency all at once. The dupes are left holding worthless tokens.

This type of scam may run for a few weeks or even a few months before it collapses. The deceived are often deeply ashamed at having been fooled by confidence men, whom they trusted solely on the basis of personality. It is nearly impossible to prosecute the crimes because cryptocurrencies elude regulation and the perpetrators typically operate in foreign countries.

In a surprisingly clever effort to educate the public about ICO scams, the Securities and Exchange Commission (SEC) launched its own slick-looking ICO website in May 2018 for HoweyCoins (www.howeycoins.com), which is an impressive, authentic-looking parody of typical ICO scam websites, complete with a jargon-packed white paper, false promises and warranties regarding government regulatory approval, and a proposal to use cryptocurrency to solve a dubious problem.[3] If you visit this website and try to purchase HoweyCoins, you are directed to a page at a Securities and Exchange Commission website (www.investor.gov/howeycoins), which informs you that you were about to fall for an ICO scam and provides more information about such activities.

It is a Piscean delusion to believe that cryptocurrencies, intangible strings of digital code with no intrinsic value, are commodities. To further confuse potential dupes, the Holy Brotherhood is running its own proprietary cryptocurrency scam by the issuance of redundant "Bitcoin futures contracts."[4] Cryptocurrencies were never intended to be traded like commodities. A lot of trouble may be avoided by rejecting the Holy Brotherhood's deceptions and using Bitcoin and altcoins as they were intended to be used, which is as a medium of peer-to-peer transaction.

Reality notwithstanding, the Feds tax Bitcoin as if it were a commodity or property. As formidable black magicians, who have the power to speak things into existence and transform one thing into another by simply saying it is so, they may consider your cryptocurrencies capital gains. Also, what they regard as a taxable event changes like the wind, so consult a qualified tax accountant or tax attorney to sort through these delusions.

Do not mistake custodial cryptocurrency exchanges for wallets. A cryptocurrency exchange is an online company that changes cryptocurrencies and sometimes fiat currency. Some popular exchanges deceptively use the term, "wallet," to describe their services. These are custodial cryptocurrency exchanges, which offer to hold your cryptocurrency for you.

Because they are traditional banking companies, they are easy for most people to use. They automatically generate addresses for you to receive cryptocurrency, however, you will never see your private key because only they have it, which means they own your cryptocurrency. Many of these exchanges are very popular with commodities traders, who think Bitcoin is a "digital commodity" and who often do not even know what a private key is. Such exchanges may provide comforting but bogus warranties and guarantees by the Securities Exchange Commission or the FDIC. But, their promises are as genuine as the money issued by a central bank. It is never safe to store cryptocurrency in an exchange or with any other third party. Using one of these services puts you firmly right back into the hands of the Holy Brotherhood.

Potentially, custodial cryptocurrency exchanges are ripening exit scams in which their operators wait for just the right moment to close up shop and disappear with the contents of their customers' "wallets." They collect your personal and financial data and, in the United States, they are subject to the same KYC-AML laws as the banks. At any time, they may use any excuse to confiscate your assets, which they will do without recourse to you. Be suspicious of anyone requesting a copy of your driver's license, social security number, legal name, or banking information to set up an account. Please, see the next chapter for information on more secure methods of exchanging cryptocurrencies.

Endeavor to understand the technology. Cryptocurrencies carry similar risks as any other technology when it is used by a person who does not understand it. Taking a little time to gain an understanding how Bitcoin works makes it easier for you to spot potential scams.

For example, there is a company whose come on is: "We store your Bitcoin in our vaults in Switzerland." Of course, Bitcoin does not need to be stored in a Swiss bank vault and, because it is intangible, this would be impossible. Moreover, you do not need someone else to "hold" your private keys for you, whether in a secret vault or anywhere else. It is absurd to anyone who understands the technology. Although the idea of millionaires putting their Bitcoins in "deep cold storage" in a Swiss vault must sound very impressive to those who do not.

Similarly, many security features offered to secure your accounts at third-party websites will not protect you. They give their customers a false sense of security and provide hackers with even more opportunities. Of course, no amount of security will protect your cryptocurrency from theft once you have handed it over to a third party.

Take extra steps to secure the privacy of your electronic devices. Use clean digital devices, free of malware and viruses. Secure your Internet connections. Use reliable VPNs (Virtual Proxy Networks) whenever accessing any cryptocurrency-related websites to thwart trackers and data collectors.

Never assume that no one is attempting to track your activities because they are, whether it is your Internet service provider, your browser, your search engine, or other third parties, whose services you voluntarily use, or nebulous intelligence firms whose business is collecting, collating, and brokering your data. No matter what you do with your Internet-connected devices, whether laptops, phones, personal assistants, or body-worn fitness devices, you are being tracked and are vulnerable to hacks, unless you take extra steps to secure your own privacy.

Be aware of volatility. In finance, the term, "volatility," refers to the varying highs and lows of the value assigned to a security. While Bitcoin is not a security, it is similarly valued against the exchange rates of fiat currencies.

The volatility of Bitcoin and several popular altcoins may be attributed to a number of causes. Organic reasons include the fact that, unlike a bank, blockchains never close, which allows for constant, uninterrupted transactions. The number and size of the transactions, also, affects its value. But, in the case of highly volatile cryptocurrencies, especially Bitcoin, the extremes are due to organized groups of speculators "pumping and dumping" them, trading them as if they were Wall Street stocks.

To cope with Bitcoin's volatility, merchants usually place a time period on a transaction. For instance, Bitcoin invoices contain an expiration, after which the proposed transaction is no longer valid. After it expires, the agreement must be renegotiated. Many altcoins, some of which are superior for transacting business, are considerably less volatile than Bitcoin.

Verify the accuracy of addresses. When transacting, it is advisable to cut and paste keys or use QR codes, then always verify the information before sending it. If you send it to the wrong address, it will just be gone. This has not been reported as a big problem, but it could happen. Usually, if you mistype a wallet address, the transaction simply cannot be processed and the funds will never leave your account.

Transact in smaller amounts rather than larger ones. It is more discreet and safer to transact with smaller amounts of cryptocurrency than larger ones.

Selling $3,000 or more in cryptocurrency may violate Federal law. Various state laws may, also, apply.[5] The relatively small number of arrests of cryptocurrency scammers made by the Feds have involved people not only transacting amounts of over $3,000, but also charging very large fees.

Bear in mind the Holy Brotherhood's Modus Operandi. Historically, when the Holy Brotherhood perceive the first vestige of a threat to their commercial interests, they react by demonizing it and anyone associated with it. They have been using the same strategy against their enemies for centuries.

Soon after the allopathic medical establishment had begun formalizing itself in the thirteenth century, the Holy Brotherhood directed its wrath at herbalists, healers, and midwives, because they presented a threat to that industry.[6] As a result, alternative medicine is highly suppressed today and practitioners risk fines and imprisonment.

According to historical researchers, in Western Europe during the Middle Ages was witches as healers threatened the burgeoning and highly lucrative sugar industry, in which the churches had an interest, by advising their patients not to consume sugar. They observed the link between sugar consumption and a range of diseases, which they had not seen before its widespread introduction to Europe.[7] The church and state targeted them for persecution while they suppressed information about the ill effects of sugar. Incidentally, its consumption is linked to common eyesight difficulties, which are now taken for granted in older people, as well as more commonly known conditions, such as diabetes, dental caries, and obesity.

Similarly, in the 1930s, when U.S. hemp growers presented strong commercial competition to the powerful petrochemical industry, they demonized this ancient medicinal plant.[8] Medical heretics were demonized along with the plant and paid a heavy toll for defying the Holy Brotherhood. Fortunately, because of easy access to accurate information about this powerful healing plant via the Internet, it is becoming decriminalized.

In accordance with their modus operandi, the Holy Brotherhood has undertaken a defamation campaign against Bitcoin and its users. They portray banking heretics as an underworld class of criminals, drug dealers, thugs, pornographers, and thieves. Modern devotees of Most Holy Death will readily perceive the parallels between anti-Bitcoin propaganda and their similar defamation of Santa Muerte and her devotees, whom they portray as gangsters, drug addicts, prostitutes, and underworld criminals. Of course, these are more examples of the Holy Brotherhood's inversions of reality, since it is difficult to imagine more perversion and criminality than that which exists among God's most faithful.

The central bankers' involvement in drug trafficking was known in the 1934, as revealed in the following statement by Congressman Louis McFadden:

> *Mr. Chairman, when the Fed was passed, the people of these United States did not perceive that a world system was being set up here which would make the savings of the American school teacher available to a narcotic-drug vendor in Acapulco.*[9]

Bitcoin is certainly no more used for criminal activity than the U.S. dollar, the Mexican peso, and the British pound are by the bankers themselves, who do business with Mexican cartels,[10] drug dealers,[11] and international terrorists with impunity.[12]

Release old ways of thinking. The illusions of the Piscean Age are still pervasive. It can be easy to fall back into old, comfortable, yet hazardous, patterns of thinking. While it may sounds comforting to believe in an unseen, fatherly authority figure who is watching over your financial interests for you, guarding your funds, keeping you safe from harm, this is a dangerous delusion and it always has been. It is important to release these old ways of thinking.

Unleash your creativity. Imagine, innovate, develop, and invent new ways of doing things with Bitcoin. Bitcoin's technology has led to many new developments. It's architecture leads to a complete restructuring of present models of communication and finance. Consider how you might do things you do now differently with Bitcoin and altcoins. How might

Bitcoin and its technology make it possible for you to do something you always wanted to do? You do not need to ask anyone's permission. Simply, do it.

Know, Dare, and Maintain Your Privacy

The Holy Brotherhood's ancient, centralized systems of power, which include religion, medicine, banking, organized commercial enterprises, guilds and universities, are all intricately woven together like a huge spider web. It is important to realize that the Holy Brotherhood and the Men of God have not improved themselves. To the contrary, they have only tightened their grip and, if they get the opportunity to openly reprise the brutality of their not-so-distant past, they will do so.

Secrecy is still a necessity for heretics because, as the Age of Pisces is in its death throes, its desperate masters are more dangerous than ever because *they know their time is short*. Remember the lessons of the past and play your cards close to your vest. Never underestimate the Holy Brotherhood's greed and arrogant sense of self-entitlement, nor the faithful's love of authority and willingness to support them even when they openly commit the most depraved acts. Let your inherent right to privacy cloak you from their diabolical machinations. *Know, dare, and maintain your privacy.*

CHAPTER 12
FREE, FUN, LOW AND NO-RISK WAYS TO GET BITCOIN AND ALTCOINS AND HOW TO USE THEM TO PROTECT YOUR PERSONAL AND FINANCIAL PRIVACY

Quickly get started securing your digital privacy, acquiring Bitcoin and altcoins with little to no risk, and using them to obtain what you want. Easily adapt conventional models of communication and transaction to Bitcoin and altcoins for anti-censorship and privacy.

Secure your Digital Privacy

Use clean electronic devices, which are free of malware, with secure connections. While this is generally good advice, it is especially necessary when you work with cryptocurrencies because they are valuable and users are targeted by thieves. But, if you take some basic steps to secure your privacy, you will be a hard target rather than a soft one. At a minimum, use a VPN, secure browsers, and secure email accounts, which employ a high level of encryption.

Virtual Proxy Networks

A Virtual Proxy Network (VPN) hides your IP address from the administrators of the websites you visit. Unless you are connecting to the Internet through a public network, a free VPN service applied to your browser should be sufficient privacy protection for your everyday activities.

Try to avoid the use of unsecured or poorly secured public networks, such as those libraries, coffee shops, airports, and hotels provide to their customers. If you must use them, invest in a subscription VPN service and apply it to your computer's settings so that it is activated immediately upon start up. Full-computer VPN services sometimes require a download. Increasingly, secure VPNs allow you to pay with Bitcoin and altcoins for even more privacy.

A VPN does not protect you from snooping by your Internet service provider or browser. It does not provide anonymity and is not sufficient to hide illegal activities. It only shields your IP address, your primary online identifier, from the sites you visit and third-party trackers.

If you need true anonymity, learn how to use Tor browser with the Tails operating system, launched from a USB flash drive. You must, also, learn how each of these work and their limitations. There are no 100% sure methods of obtaining anonymity, even with Tor and Tails, especially if you are doing something that is likely to garner unwanted attention.

Even if you never visit a cryptocurrency website, your activities are being tracked and this information is being collected and sold, not just to government agencies, but to common criminals on the black market. Therefore, a free VPN is basic protection for all general use of the Internet so you do not leave the most obvious traces of your identity at every website you visit.

Secure Browsers

Secure browsers are those that include privacy features to prevent third parties from easily identifying and tracking you. They may, also, prevent the execution of malicious scripts embedded in web pages you visit.

Epic Privacy Browser (epicbrowser.com) is a secure, Chromium-based browser, which prevents tracking by third parties and features an automatic, built-in ad blocker, which blocks advertisements and malicious scripts. It is equipped with a built-in VPN, which is easily enabled by clicking an icon in the upper right-hand corner. Whenever you close Epic browser, it will remember your VPN setting, but dump cookies and other tracking data. It is one of the most secure options for everyday browsing and is easy to download and begin using right away.

Opera (opera.com) is an attractive, easy-to-use, full-featured browser, which is good for every day use. It provides a built-in VPN, which must be activated in the browser's settings or toolbar. Opera is the first popular browser to address the problem of malicious cryptocurrency mining scripts, which surreptitiously use website visitors' CPU (Central Processing Unit) or GPU (Graphics Processing Unit) power to mine cryptocurrencies. It, also, features a built-in ad blocker, which must be activated in its settings.

Tor (torproject.org) is a browser for special occasions when you need to take extra steps to protect your identity. Tor provides access to the "Dark Web," which are websites that are intentionally hidden from popular search engines. It is used by political activists, journalists, whistleblowers, intelligence agencies, common criminals, tech enthusiasts, and privacy seekers. For the highest degree of anonymity, use Tor with the Tails operating system launched from a USB flash drive, according to the instructions at the Tails website (tails.boum.org). Alternatively, pair Tor with Whonix Anonymous Operating System (whonix.org).

Tor is not a silver bullet against privacy intrusions. Always pair it with a compatible operating system, otherwise, it is easy for someone to trace your activities back to your device. Study how Tor works at torproject.org before attempting to use it for anonymity. There are potential hazards at entry and exit nodes of the Tor system. Once in the system, it is important not to provide identifying information, even one time. For instance, never use Tor browser to visit a custodial cryptocurrency exchange where you provide your real name, phone number, banking, or other revealing information.

Ad Blockers

When using popular browsers, apply and use ad blockers to prevent the execution of malicious scripts and redirects of your web browser to pages, which may contain malware or viruses. Do not disable ad blockers for any reason. Sometimes websites will try to guilt you into disabling them. Don't fall for it. If a site is unusable because it will not allow you to use your ad blockers, find another website that offers what you want without invading your privacy. Find an ad blocker add-on or extension for any browser by performing a web search for: A*d blocker for [name of the browser]*.

Secure Email

When working with cryptocurrencies, do not use popular email services, which store your private information without encryption, scan your emails, sell your personal data, and expose you to security breaches and subpoenas. Instead, use secure email services, which provide layers of encryption, do not sell your data, and will not hand over your unencrypted data to government agents. Tutamail or Tutanota (tutanota.com), based in Germany, and Proton Mail (protonmail.com), based in Switzerland, are examples of innovative, easy-to-use, secure, web-based email services that incorporate high levels of encryption. Both offer free and paid services for personal and business use. Proton Mail, also, provides a VPN service, ProtonVPN, which may be activated in its settings. Proton Mail is a favorite of Dark Web users.

To Simplify

Use Epic (epicbrowser.com) when you need a no-fuss, private browser with a built-in VPN and ad blocker. For secure email, use Tutamail/Tutanota (tutanota.com), which is so secure that it does not require the use of a VPN. These are the simplest options for everyday online security.

Bitcoin and Altcoin Favorites

There are many cryptocurrencies to choose from, and new altcoins constantly emerge. Each of the author's favorites, below, have a measure of acceptance, a well-developed merchant interface, a philosophy of privacy and self-autonomy, and did not begin as an Initial Coin Offering (ICO).

Bitcoin (bitcoin.org): The pioneering, decentralized, P2P cryptocurrency. It possesses the advantages of its brand name and the greatest acceptance as a payment option, which presently makes Bitcoin indispensable.

Litecoin (litecoin.org): A Bitcoin clone, forked from the Bitcoin Core client. If Bitcoin were gold, then Litecoin might be silver. It is increasingly accepted by merchants, and its transaction fees are low.

Dogecoin (dogecoin.com): Begun as a joke, it is fun and inexpensive to use. Dogecoin is represented by its iconic, cute dog, a Shiba Inu, and is commonly used for Internet tipping, donations, and small charitable contributions. It is favored by many merchants and web denizens.

Dash (dash.org): Formerly known as "DarkCoin," it is a fork of Litecoin, which includes a "private send" feature. Dash is enjoying increasing adoption by merchants and privacy lovers.

BitcoinCash (bitcoincash.org): A 2017 fork of Bitcoin, which some users claim is the "true" Bitcoin because it aims to uphold the ideal of maximum decentralization.

BitcoinGold (bitcoingold.org): Another 2017 fork of Bitcoin, which has gained some traction and has a similar philosophy to that of BitcoinCash.

Monero (getmonero.org): Built for anonymity and freedom from censorship, Monero best represents the true mission of Bitcoin and the Cypherpunk philosophy of using digital encryption to secure individual liberty through privacy. While Bitcoin is pseudonymous, Monero is truly anonymous and provides maximum privacy. The creation of a Monero account generates a "viewing code," which allows only you to view your account balance and transaction history. Unlike other cryptocurrencies, establishing just one Monero account can protect your anonymity and security well, as long as you do not compromise it.

Monero's development team is involved in a number of exciting, cutting-edge privacy innovations, including Kovri (getkovri.org), an I2P-based onion router network, similar to Tor. Also, the Monero team and its community have proven themselves highly resistant to centralization.[1] Monero is especially valued by those who appreciate its unwavering commitment to free commerce and free speech, which is why it is the preferred cryptocurrency of the Dark Web and this author.

Bitcoin does not require your fidelity. Cryptocurrencies do not make declarations, such as "Thou shalt have no other cryptocurrency before me." As a banking heretic, you are free to use or not use any of them you choose and to change your mind at any time. Bitcoin and altcoins are constantly emerging and evolving. Some developments are improvements, but not all. No cryptocurrency is infallible. Each is simply a system of transaction. The more altcoins in use, the safer the big, original idea of Bitcoin is. Pandora has opened the proverbial box, but it is we who have been set free because the Holy Brotherhood cannot put Bitcoin and all its infinitely replicating clones back inside.

Get a Wallet

"Wallet" is a metaphor for how you store your cryptocurrency account information. To get a community-recommended wallet for any cryptocurrency, visit its official website. As you become more familiar with Bitcoin and altcoins, you will use different types of wallets.

Hot wallets are those that connect to the Internet. These are generally for storing a small amount of funds, which you plan on spending soon. Cold wallets are for longer term storage of funds, which is humorously called "cold storage." Good cold wallets are made with no Internet connection so that the private key to your account cannot be stolen. Once private key information is connected to the Internet, it becomes "hot."

Some wallets function as only hot or cold, while others perform as both, allowing you to import the private keys of cold accounts, so you can spend the funds.

Hot Wallets

Web wallets: These are provided by individual websites and are easy to use for making online purchases. The most secure web wallets provide client-side services and do not have access to your private keys. Some offer their services as add-ons and extensions for popular browsers. Many web wallets, also, provide corresponding smartphone apps.

Smartphone Apps: These are the most user-friendly option for making purchases anywhere by scanning a QR code. They are not especially secure and should only be loaded with relatively small amounts of cryptocurrency.

Hot-cold Wallets

Desktop wallets: These are software programs you download to your computer. Generally, they are more secure than web wallets and smartphone apps. Electrum (electrum.org) and Bitcoin Armory (bitcoinarmory.com) are examples of desktop wallets for Bitcoin that allow the creation of both hot and cold wallets. Also, the keys to any cold Bitcoin accounts may be imported into these programs.

Hardware wallets: While these are very secure for long-term cold storage, they are, also, easy to use for sending payments. Hardware wallets, such as Ledger (ledger.com), Trezor (trezor.io), KeepKey (keepkey.com), are primarily kept offline, but may be plugged into a USB port. Most support multiple cryptocurrencies or may be modified to do so. Presently, they retail at around $100 to $230.

Cold Wallets

Paper wallets: These are simply the public and private key pairs to an account printed or written on a piece of paper. They are inexpensive, practical, and very secure as long as you do not lose the paper or expose the information on it to theft. Paper wallets are created offline and stay offline until you are ready to import the funds into a hot wallet.

The following websites provide paper wallet generators:

Bit Address (bitaddress.org): For Bitcoin only. After you navigate to the web page, disable your wi-fi connection, then generate your account keys.

(https://github.com/pointbiz/bitaddress.org)

Monero Address (moneroaddress.org): Create a paper wallet for Monero or Aeon, similar to the above.

Wallet Generator (walletgenerator.net): Similar to the above, but it generates cold wallet keys for many cryptocurrencies. To very securely generate a paper wallet, download the Wallet Generator's Github files and use them offline, as described at:

https://github.com/MichaelMure/WalletGenerator.net.

Removable storage devices: USB flash drives and writable CD/DVDs may be used to store information related to your paper wallets. Use them in addition to or instead of paper. For stronger security, encrypt the information.

Whenever you make cold wallets this way, save the key pairs together. Make notes, such as the creation date and the purpose of the account. Before you use a new address, verify that it was created on its respective blockchain. Locate any cryptocurrency's public ledger by performing a web search for its name combined with the term "blockchain," for example: "Bitcoin Blockchain," "Monero Blockchain," "Aeon Blockchain," and so on. There you should see your new account and be able to monitor your transactions.

When you are ready to spend your cold stored funds, sweep the contents of the entire account into a hot-cold wallet using the private key code. Importing and sweeping are similar acts, but sweeping means to transfer all the funds from a cold account out to new accounts for security purposes. Once you have emptied or "swept" the contents of a cold account, you may spend a portion of it from your hot wallet and send any remainder to a new cold account for continued safekeeping.

QR Codes

Bitcoin and altcoin transactions may be quickly performed by scanning a QR ("Quick Response") code with a smartphone app. To find the QR code for a cryptocurrency key, go to duckduckgo.com and insert your request into the search box, as follows: *QR code xxxxxxxxx.*

Get Bitcoin and Altcoins

When many people first hear about Bitcoin, they want to buy some. Sometimes they provide their banking information to strangers and purchase thousands of dollars worth at a custodial cryptocurrency exchange. But this is not the best way to get cryptocurrency because it risks your cash and your privacy. Fortunately, there are many ways to obtain Bitcoin and altcoins without such risks.

Faucets

Faucets are websites that "drip" a little bit of cryptocurrency. Users must get their own wallets, then play trivia and other games to receive a payout. If you enjoy them, it is possible to get anywhere from a few dollars to more than a hundred dollars worth in a year, depending on cryptocurrency volatility.

Since a Bitcoin is worth far more than a dollar, you will receive a fraction of a Bitcoin, called a "Satoshi." Named for Bitcoin's founder, a Satoshi is the smallest denomination of a Bitcoin, which is one hundred millionth of a single Bitcoin (0.00000001 BTC).

1 Satoshi	= 0.00000001 BTC
10 Satoshis	= 0.00000010 BTC
100 Satoshis	= 0.00000100 BTC
1,000 Satoshis	= 0.00001000 BTC
10,000 Satoshis	= 0.00010000 BTC
100,000 Satoshis	= 0.00100000 BTC
1,000,000 Satoshis	= 0.01000000 BTC
10,000,000 Satoshis	= 0.10000000 BTC
100,000,000 Satoshis	= 1.00000000 BTC

Similarly, Litecoin is expressed in "Litoshis." Other popular altcoins are expressed in decimal denominations.

Some faucets are funded by donations from the cryptocurrency community. Others are commercial enterprises that rely on income from advertising. Frequently, faucets unexpectedly "dry up" or make undesirable changes to their policies. Not all of them are safe from malware. They can be targets for hackers. Even if a faucet's main page appears safe, it may contain malicious redirects that lead you to infected sites. In fact, quite a few faucets are scams to be avoided.

The following are examples of popular faucets that currently have a good reputation:

Satoshi Quiz: satoshiquiz.com (also, available as a smartphone app)
Free Bitcoin: freebitco.in
Free Dogecoin: freedoge.co.in
Moon Bitcoin: moonbit.co.in
Moon Dogecoin: moondoge.co.in
Moon Litecoin: moonliteco.in
Moon Dash: moondash.co.in

Before experimenting with any new faucets, set a restore point on your computer and back up important files. In an attempt to deter fraud, most faucets will not allow you to use a VPN at their websites, so you will have to disable your VPN. Still, use an ad blocker and do not disable it when requested.

Cryptocurrency Mining

A miner of cryptocurrencies is one who participates in processing transactions as they appear on a blockchain. Mining requires special hardware setups, called "mining rigs," which are very fast computer processors. They get hot enough to heat up a large room and require a lot of electricity, so most mining is performed in places with cooler year-round temperatures and cheaper electricity.

Because of the high costs of equipment and electricity, early miners reported financial losses. To manage the financial risks, cloud mining emerged as a solution in which a group of users share the costs and profits of mining. CPU/GPU mining, in which users participate in cloud mining using the processing power of their personal computers is low-risk but not especially lucrative for participants. Although it is possible to provide your computer's bandwidth to a website for mining in return for a small amount of cryptocurrency. Some faucets offer a CPU mining option.

Cryptocurrency Mining Scripts

Cryptocurrency mining scripts are an alternative to pay-per-click advertising programs for monetizing websites. A mining script uses your visitors' CPU power for cryptocurrency mining. It is important to disclose the use of such a script on your website because it may appear suspicious to visitors who are aware of it, since these scripts can be used maliciously. The scripts themselves are a target for hackers, but high traffic websites may still find them beneficial.[2] To obtain a script, perform a web search for: *Cryptocurrency mining script.*

Accept Bitcoin and Altcoins as Payment

The best way to get larger amounts of Bitcoin and altcoins is in exchange for your products and services. If you already do business online, simply announce that you now accept cryptocurrencies. If you do a very high volume of transactions, you may prefer to accept payment by means of a payment processor integrated into your website's shopping cart system.

If you do a lower volume, accept cryptocurrency payments in peer-to-peer fashion, without a third-party, just as you would accept cash. In this case, your customer places an order by email. Then, you reply with an invoice, which provides a unique receiving address. Once the customer sends the payment to that address, you fulfill the order.

Independent Contractors

Use your skills to earn Bitcoin as an independent contractor. Some websites, such as XBTFreelancer (xbtfreelancer.com), function as brokers that pair freelancers with Bitcoin-paying clients. Your pay is held in an escrow account, which is only released by the client upon satisfactory completion of the job. These sites employ optional, third-party overseers who arbitrate in case of a dispute, however, the service providers must pay the costs of arbitration.

Other job websites, such as Coinality (coinality.com), are merely classified listings at which you may find a client with whom to do business peer-to-peer. Similar listings may be found at Reddit, Craigslist, and other classified ad websites.

If you are already a freelancer with a base of clients, let them know that you prefer to be paid in Bitcoin or altcoins and give them incentives for doing so.

A P2P Marketplace

Open Bazaar (openbazaar.org) answers the cry for freedom from restrictive, centralized marketplace platforms. It is the Bitcoin version of Ebay and Etsy. An article at the *OpenBazaar Blog*, entitled "OpenBazaar Seller Guide–What To Expect in This Decentralized Marketplace," describes how to use this peer-to-peer buying and selling platform for Bitcoin and select altcoins.[3] There are no restrictions on products or services that can be listed at OpenBazaar, as long as they are not illegal.

OpenBazaar does not charge fees for either listing or selling your items or services, they do not collect your data, and there is no censorship. But, you are responsible for running your own store on the network. You must download OpenBazaar's peer-to-peer, desktop software and run it from your own computer or from a cloud service. Once you create your store, save a copy of it on removable storage device for future reinstallation.

Crowdfunding

Crowdfunding is another form of alternative finance that emerged after the 2008 banking debacle. It is a means to obtain funds for a project or entrepreneurial venture. Typically, it is used to fund art, music, films, blogs, vlogs, podcasts, and other creative projects. This is how it typically works: You propose a project; your supporters donate and pledge funds for it; and an online, third-party platform facilitates and oversees the process.

Crowdfunding with Bitcoin and altcoins provides relief from the third parties, including the private platforms themselves, banks, and payment processors, which may disrupt your business, censor you, and compromise your personal and financial privacy. Donation-based and reward-based crowdfunding are two methods of conducting business that may especially benefit witches and occultists.

Donations

Donation-based crowdfunding lets people give to you freely with no expectations that you will give them something specific in return. Making and accepting donations peer-to-peer is easy with Bitcoin and altcoins.

If you have a website, blog, vlog, digital or print publication, or email list simply add your public cryptocurrency addresses and corresponding QR codes to them. Then, ask your readers, viewers, customers, fans, and friends to support your work.

Reward-based Crowdfunding

Reward-based crowdfunding allows patrons to pledge monetary support for your work, usually on an ongoing basis, in exchange for a non-monetary reward. Patreon (patreon.com) is a fine example of the patronage model. Its business concept is based on patronage of the arts, as was done by monarchs and other affluent art lovers in Medieval and

Renaissance Europe. When talented artists were financially supported by those who appreciated their work, they were able to fully devote themselves to it, which greatly elevated the quality of their creations. Michelangelo, DaVinci, Beethoven, and Shakespeare earned, at least, part of their incomes from the support of patrons.

Patreon is a third-party platform that facilitates transactions using banks, credit cards, and payment processors. Patrons agree to monthly recurring charges based on the level of support they choose to give the creator in exchange for set rewards.

The following is an example:

Level 1: $1 Gratitude only or a small gift.

Level 2: $5 Discounts on merchandise; digital gifts; and any rewards on the lowest level.

Level 3: 25 Access to exclusive materials; plus all the rewards of the lower levels.

Level 4: $100 More and nicer rewards; physical merchandise and gifts; special services; plus all the rewards of the lower levels.

Level 5: $250 Credits on the film or musical release; membership in a special group; plus all the rewards of the lower levels.

The creator may, also, set a monetary goal for the purpose of completing a project, such as a film, or funding for equipment. An example of a goal is: $5,000 to purchase studio equipment. When this goal is met, patrons receive a big reward.

Examples of rewards for patrons: Exclusive blog posts; audio files; video files; addition to an email list; educational courses; monthly news letters; giveaways; discounts; unique desktop wallpaper; handmade jewelry, knickknacks, poppets, and other arts and crafts; custom astrological charts; monthly tarot readings; exclusive recipes for potions; exclusive spells and rituals; counseling sessions; access to live streams; and merchandise, such as stickers, bumper stickers; mugs, ornaments, key chains, pendants, pillows, hats, magnets, T-shirts, embroidered patches, and custom tarot and oracle decks. Creators may obtain custom merchandise from on-demand merchandise websites.

How to adapt this idea to Bitcoin and altcoins: Eliminate all the major third-parties by managing your own patrons and rewards, and accepting Bitcoin and altcoins. Since it is not possible to automatically deduct funds from a Bitcoin or altcoin account, you must email an invoice to your patrons each month, providing them with a unique cryptocurrency account address for the payment. Give extra benefits to your patrons who pay with cryptocurrencies and reward them for their loyalty when they continuously support you after six months, one year, two years, and so on.

While you do not have the conveniences provided by the third party platform and its payment processors, neither is there any monitoring of your activities or potential interference. You cannot be shut down this way.

Memberships

Similarly, Bitcoin and altcoins may be adapted as P2P payment for membership programs. Memberships differ from donations and patronage in that members enjoy participation and sometimes collaboration in a project.

Many third-party platforms allow for private conversations, such as Yahoogroups (groups.yahoo.com), Wordpress (wordpress.com), and Slack (slack.com). Be aware that these third parties may keep your discussions private from outside observers, but will not prevent those who operate the platform from seeing them.

Privately owned and operated forums, which you host at your own website, typically feature members-only areas. If you require a little more privacy for your discussions, then this may be a better option. For maximum privacy, look at new developments in open source, peer-to-peer communication and collaboration software. Your membership program may, also, include participation in a private email list. Encourage your members to use secure email services for maximum privacy.

Membership dues may be collected monthly, annually, or for a lifetime. You may still run a conventionally managed and funded membership website or group while accepting Bitcoin and altcoins for your own self-managed program. Give extra rewards to your Bitcoin and altcoin-using members and cultivate their loyalty.

Subscriptions

Just as magazines and newspapers offer subscriptions to their publications and accept payments by check or money order, you may accept P2P payments of Bitcoin and altcoins for subscriptions to either digital or print materials. Accept a set amount in Bitcoin for 6 months or a year of your publication, access to podcasts, or video content. When the subscription expires, offer your customer an option to renew and incentives to do so using cryptocurrency.

Privacy for Anti-censorship and Covens

There are extra considerations when using the patronage, membership, or subscription model for anti-censorship and privacy for covens and virtual societies. When communicating material that might be controversial or when communications between members of an organization require secrecy, use encrypted email to deliver content to your patrons, members, and subscribers.

You can achieve a high degree of privacy by using secure email to deliver links to your files hosted at third-parties that have a good record of protecting privacy. You may, also, directly deliver encrypted files to your supporters and colleagues using AES-256 encryption and strong passwords. You must give them their decryption passwords by another means, for instance, by text, phone, conventional mail, or in person. Additionally, accepting payments in Monero provides maximum protection for you and your associates.

Bitcoin ATMs

Several companies manufacture Bitcoin ATMs (Bitcoin-Automatic Teller Machines), which have different features and requirements. Most permit only one-way transactions, fiat to Bitcoin; but others convert currency both ways. Some convert altcoins, too.[4] Some Bitcoin ATMs require a scan of your driver's license, a fingerprint, hand scan, a photograph of you, or mobile phone number, although some require no such identity disclosures. Some of them employ surveillance cameras similar to conventional ATMs. They typically charge expensive fees of 5 to 12%. If they do not require personal or identity information from you, they are a relatively safe way to obtain a tiny amount of cryptocurrency and experience the novelty of a Bitcoin ATM. To find current locations of Bitcoin ATMs, see a map of ATMs at Coin ATM Trader (coinatmradar.com) or perform a web search for: *Bitcoin ATMs*.

Cryptocurrency Exchanges

Cryptocurrency exchanges are companies, software, and services, which facilitate the exchange of cryptocurrencies. Presently, there are three main categories of exchanges: Centralized custodial; centralized non-custodial; and decentralized peer-to-peer.

Centralized Exchanges

There are two main types of centralized exchanges: Custodial and non-custodial. Custodial exchanges are those that hold your cryptocurrency for you, denying you access to the private keys.

In contrast, non-custodial exchanges are centralized online companies, which exchange one cryptocurrency for another or cryptocurrency for fiat, without taking control of your private keys.

Each represents different degrees of centralized control. The former is structured like conventional stock exchange websites and is rife with fraud, while the latter is considerably less dangerous. Although neither are ideal.

Centralized, Custodial Exchanges

Centralized, custodial exchanges are popular among many newcomers to Bitcoin because they operate like traditional banking services. Undoubtedly, the fact that they look and function just like conventional trading platforms adds to the mass delusion that cryptocurrencies are commodities. Additionally, they are integrated with popular third-party shopping cart services for online merchants. They enjoy unquestioned acceptance among those who have only superficial knowledge of Bitcoin.

Within custodial exchanges, Bitcoin, altcoins, and fiat currency may be exchanged by the company that owns the exchange. For instance, if you exchange Bitcoin for Litecoin, there is no transaction recorded on either of these blockchain's ledgers. It is all done on the company's own servers. Users may purchase cryptocurrencies from the exchange with their bank accounts or deposit dollars from the exchange directly into their own bank accounts. There may be a sizable fee for exchanges and withdrawals.

Custodial exchanges require you to establish an account, create a password, and usually require some form of authentication to login. They engender in their users a false sense of security, while depriving them of the private keys to their accounts. At some point, the company may require identity verification and banking information, including your bank account, social security number, and a copy of your driver's license, even if you only exchange cryptocurrencies. Although they may wait until you have deposited a tidy sum, effectively holding your funds hostage until you provide whatever information they demand.

They place similar restraints on their users as other centralized third-parties do. If they disapprove of the business you are in, if they dislike something you did, if the laws in your state change, or if they deem you suspicious, they may close the account and take your funds. Like other centralized corporations, they keep databases that include your identity and records of your transactions, most likely stored without encryption, which are vulnerable to internal fraud and hacking, and subject to scrutiny by the Feds.

Mt. Gox, Bitconnect, and Coinbase are examples of centralized, custodial exchanges. Such companies sometimes use the terms, "wallet" and "hosted wallet," to describe their services.[5]

Mt. Gox was a centralized, custodial exchange based in Tokyo, which operated from 2010 to 2014. It performed an apparently respectable service for its customers until one day, at the height of its productivity, it was simply gone. In the several months between 2013 to 2014, right before it abruptly closed up shop, it handled 70% of worldwide exchanges involving Bitcoin.[6] While the company initially reported that they had been the victims of theft, in 2015, Japanese law enforcement agents charged the CEO with embezzling £1.7m ($2.6 million USD).[7]

Bitconnect was a short-lived centralized, custodial exchange, which began as an ICO. Experienced investors recognized it as a High-yield Investment Program (HYIP) and plaintiffs, in a U.S. lawsuit filed against the company, alleged it to be a Ponzi scheme.[8] In January 2018, Bitconnect proved to be a very large exit scam. A court in Kentucky froze the company's cryptocurrency assets,[9] however, it is highly unlikely that the stolen funds will be recovered.

Coinbase is a centralized, custodial exchange, founded in 2012 and headquartered in San Francisco,[10] which offers the service of placing your cryptocurrency in cold storage for you.[11] According to a blog post by their CEO in the United Kingdom, they have a "commitment to making cryptocurrency accessible to everyone."[12] While this may sound like a generous promise, if the faithful avariciously view Bitcoin as a commodity, they will likely be taught another lesson by the Holy Brotherhood.

Coinbase users have already been subjected to a blanket subpoena of their data.[13] After a legal scuffle, in which the Feds demanded the information of every single user in Coinbase's database, a summons was issued by a Federal judge to limit the scope to any transaction amounting to a minimum of $20,000.[14] The IRS wants this information because it fears that Bitcoin users are making large sums of money, which they are not reporting to the Feds. In fact, entire fortunes have been lost investing in and trading cryptocurrencies via custodial, cryptocurrency exchanges.

Centralized, Non-custodial Exchanges

There are a few centralized, non-custodial exchanges, sometimes called "instant cryptocurrency exchanges," which exchange one cryptocurrency for another without requiring you to give up your private keys or any identifying information. You simply choose the cryptocurrencies and the amounts you want to exchange at their website and provide your own account addresses.

Changelly (changelly.com), a company founded in 2013 and headquartered in Prague, is an example of such an exchange. There are similar instant exchanges, all of which apply different rates and fees, so shop and compare. To find them, perform a web search for: *Instant cryptocurrency exchange.*

Decentralized, Peer-to-Peer Exchanges

The most secure and private exchanges are decentralized, peer-to-peer, and ultimately person-to-person, with no third-party. Highly decentralized systems are best suited to working with cryptocurrencies because they share the same structure. There are relatively few P2P exchanges because they require more effort to use, however, they are preferred by those who understand the technology.

Peer to Peer Vs. Central Server System

Peer-to-peer (P2P) systems allow an individual to directly connect to every other peer sharing a network. No matter who enters or leaves the network, it remains. There is no one point of attack, and no repository of data is stored in any one place.

By contrast, the conventional Internet works on a hierarchical system, with a central server at the core, through which all communication flows. If the central server disappears, the entire network goes down. This central point stores every user's data and is subject to attack, which compromises the security of every user on the network.

Bisq (bisq.network and github.com/bisq-network/bisq-desktop/wiki), formerly known as "Bitsquare," is a decentralized, P2P exchange. Once you download their software, you are able to connect with every other user on the network and make or accept offers.

Local Bitcoins (localbitcoins.com) and similar sites, such as Local Monero (localmonero.co), and LiteCoin Local (litecoinlocal.net), and Local Bitcoin Cash (localbitcoincash.org), allow users to connect with local cryptocurrency users who are willing to make an exchange.. The site uses a rating system similar to that of popular online merchant websites.

As more people begin using cryptocurrencies, it may become easier to ask a friend or acquaintance to make an exchange with you. This may already be a possibility in some areas. If you want to make an exchange, ask around.

Presently, blockchain communities are cooperating to devise ways of directly exchanging cryptocurrencies between two different blockchains. Decentralized methods of exchange, sometimes abbreviated as "DEX," are still under construction. To find the latest developments, perform a web search for "decentralized exchange" or "P2P exchange."

It is ideal to receive and spend cryptocurrency without the need to exchange it. Bitcoin is meant to be used as a medium of transaction, apart from the Holy Brotherhood's system. But, when you must exchange one currency for another, decentralized, peer-to-peer methods are best.

CHAPTER 13
PRIVACY AND THE "RIGHT TO BE LET ALONE"

Privacy is the habit of keeping secrets. Traditionally, witches have always dealt with our most powerful and dangerous enemies by working in secrecy. In the darkest hours of the night, we create and reshape the world around us. To this objective, digital encryption and cryptocurrency present a means of accomplishing our goals while concealing our activities from those who would cause us harm. By means of sophisticated digital encryption and Bitcoin, we may now communicate and conduct business privately, one-to-one, just as our ancestors did in the past. These high-tech tools return our natural power to us and remove us from the domain of our historical enemies.

Privacy is the Law

You have been taught to obey the law. Furthermore, there are civil and criminal penalties for not doing so. Fortunately, there is a body of laws summarizing your right to maintain privacy, and, thereby, protect all your other liberties.

U.S. jurisprudence sets forth the inherent "right to be let alone," originally described by the famous nineteenth-century jurist from Michigan, Thomas M. Cooley, in his book on tort laws, *A Treaties on the Law of Torts or the Wrongs Which Arise Independent of Contract*, published in 1879.[1] It was expanded upon in an article written in 1891, entitled "The Right to Privacy," by two attorneys, Samuel D. Warren and Louis D. Brandeis, whose concern was the loss of privacy in correlation with the development of new technology, such as cameras and recording devices.[2] The work of these attorneys has become the basis for modern privacy law in the United States.

Cooley's legal assertion of the right to be let alone is mostly based on the Fourth Amendment, which defines your right to privacy of your personal effects and documents; and the Fifth Amendment, which defines your right to be silent. Of course, men's laws cannot give you the right to privacy, nor can they take it away, because it is an inherent right. No matter where you were born, from the moment you took your first breath, the right to be let alone has been yours.

Interestingly, privacy laws are rarely invoked in most court cases anywhere in the United States, despite the fact that it is the primary right to be let alone, which is the very first that is violated in the commission of nearly every other crime, whether it is a crime against property or one's person. For instance, it is impossible for a criminal to commit vandalism, identity fraud, financial fraud, assault, or battery, without first violating their target's privacy. When you protect your privacy, you prevent crimes against yourself.

Exercising Your "Right to Be Let Alone"

Consider all the ways the Holy Brotherhood violates your right to be let alone. It is their purpose for every personal identifier they issue to you, the most personal information they demand from you, and every license and registration they require you to obtain. The Holy Brotherhood's financial system is a mass surveillance operation, conducted for total control over your life.

The right to be let alone is your right to keep your personal effects in your own possession and under your own control. It is your right to be free of molestation. It is your natural right to the fruits of your own labor, that which you create with your mind and body. It is your right to control information about yourself and to keep inherently private matters out of the hands of your enemies. It is your absolute, inherent right not to be hounded, tracked, and surveilled by anyone, least of all by the descendants of the same violent degenerates who have pursued witches and occultist to our deaths for millennia.

But, privacy laws themselves do not protect you, they only state your inherent right to protect yourself. Furthermore, no one but you can protect your privacy. For many, coming to this realization is part of the paradigm shift that must be made to benefit from the new age.

Secrecy is Power

In popular culture, the masses are bombarded with the message that keeping secrets is unhealthy and burdensome. They have made "secrecy" a dirty word, with connotations of shame and crime. While telling secrets about yourself may be emotionally gratifying, at least in the moment, it makes you vulnerable and puts power into the hands of people who may be your enemies. Often, you do not realize that someone is an enemy until they have already obtained personal information about you. But, when you keep your secrets, you retain your power.

Furthermore, you are powerful when you dare to act independently, without anyone else's approval or permission, which you are free to do when you exercise your inherent and legal right to privacy. The right to privacy is fundamental to liberty because it protects every other right you have. So, learn to keep secrets and keep them well. Never compromise your power, security, safety, or liberty by giving in to the temptation to give away secrets about yourself.

Silence is Golden

Practice using the privacy tools described in this book, such as encryption, secure browsers, secure email communications, and peer-to-peer transactions with Bitcoin and altcoins. Whenever you can, use cryptocurrencies and keep your personal and financial information out of centralized databases.

When dealing with others, establish an attitude of mental imperviousness and form a habit of being silent and self-defensively deceptive whenever your inherent right to privacy is endangered. With the exception of specific legal circumstances, you do not owe anyone an answer to a question. Plan ahead for situations in which you might be subject to intrusions on your privacy. Rehearse keeping silent and, when there is no other option, deceiving your inquisitors.

For witches, silence means life. It is in silence that our kind has survived all these many years.

Know, dare, and remain silent!

GLOSSARY

AES-256 Encryption: A high standard of encryption, which remains unbroken. AES is an acronym for "Advanced Encryption Standard." There are many programs for encrypting and decrypting files using AES-256, including, 7-zip (7-zip.org), which is supplied as a standard app on some newer versions of Windows, and may be downloaded from their website.

Altcoin: A cryptocurrency other than Bitcoin, such as Litecoin, Dogecoin, or Monero.

Atheist: One who does not believe in or does not acknowledge the existence of the god of the Sinister Trinity, also, referred to as the Supreme Being.

Bank: An institution that receives deposits, makes loans, exchanges currencies, provides other financial services, and guards assets in its vaults. A store of coins, such as a "piggy bank."

Bank notes: IOUs issued by a central bank and circulated as currency. Usury-based currency.

Bill of Rights: An addendum to the U.S. Constitution, which describes a few fundamental human rights, which are inherent to all people. The two most closely related to the right to privacy are:

> **Fourth Amendment:** "The right of the people to be secure in their persons, houses, papers, and effects, against unreasonable searches

and seizures, shall not be violated, and no warrants shall issue, but upon probable cause, supported by oath or affirmation, and particularly describing the place to be searched, and the persons or things to be seized."

It includes digital privacy and the right not to be surveilled, however, this is a right you must take into your own hands. State-of-the-art digital encryption is a powerful tool in removing the opportunity for either a state agent or a common criminal to violate these rights by simply removing their ability to do so.

Fifth Amendment: "No person shall be subject, except in cases of impeachment, to more than one punishment or trial for the same offense; nor shall be compelled to be a witness against himself; nor be deprived of life, liberty, or property, without due process of law; nor be obliged to relinquish his property, where it may be necessary for public use, without just compensation."

Commonly referred to as "the right to remain silent," it applies to those who are under arrest or being held to account for a suspected crime, however, in its broader implications, it represents the right of the individual to keep silent, to maintain privacy, and keep secrets. You have no obligation to tell what you know, to out yourself, or tattle on yourself. There is neither a legal nor social obligation to do so. This legal right is derived from English Common Law and is found in some form in most English-speaking countries. It's another way to say: *Silence is golden!*

Bitcoin: The first decentralized, peer-to-peer cryptocurrency.

Blockchain: The digital ledger of Bitcoin or other cryptocurrencies. (See: Developer Documentation https://bitcoin.org/en/developer-documentation, also: https://github.com/keeshux/basic-blockchain-programming)

Censorship: The act of censoring or acting like a censor. The word, "censor," is based on the office of a Roman magistrate of the 5th century B.C, whose office was to conduct a census, and oversee public morals, public finances, and public works. Censorship may be illegal or legal in nature. Most often, the term is applied to the suppression of speech in any form by a government or its agents, which is against the law in many countries. Since governments operate behind a front of private corporations, the definition may be extended to private entities, especially when governments circumvent legal restrictions by delegating their authority to censor to such entities, including financial corporations, corporate owned media, and social media monopolies. Forms of legal censorship include religious censorship, creative censorship, Internet censorship, military censorship, moral censorship, corporate censorship, nearly all forms of private censorship, and self-censorship.

Central Bank: A private banking cartel, which parasitically positions itself within a government, then issues IOUs in the guise of currency, which it lends the host government at usurious interest rates. The citizens of these nations then must pay taxes to pay the interest on the national debt, which is mathematically impossible to pay off, thus creating rising inflation. The purpose of a central bank is to surreptitiously deprive nations and individuals of their resources, dispossess, and financially enslave them.

Centralization: The concentration of power and authority in the hands of a single individual or small group.

Cryptography: Secret information or secret writing, the study of it, and the rendering of information into a secret code, which one must have the decoding key to decipher. The modern binary cryptography used in digital cryptography may be traced to the famous cipher of Roger Bacon.

Currency: A medium of transaction and exchange.

Dark Web: Websites accessed through Tor browser, which are not indexed on the World Wide Web.

Decentralization: The distribution of information, activities or powers among individuals. The opposite of centralization and monopolization.

Double-entry accounting (double-entry bookkeeping): A system of accounting in which every transaction is recorded twice, as a debit and a credit. It was first described in the appendix of a book, *The Book on the Art of Trading,* by Benedikt Kotruljevićin 1458. The original title of his book is *Libro de l'Arte de la Mercatura.* When it is used honestly, it is a good method of keeping track of a company's assets and liabilities, however, when it is coupled with fractional reserve lending and the sale and resale of debt as securities, it is as a tool of deception and fraud.

Encryption: Information in the form of code. The act of encoding information so it cannot be intercepted by unauthorized parties. The opposite of decryption.

Exchange: A company, software, or service that facilitates changing one currency for another.

Federal Reserve Bank: The private central bank that manufactures and lends usury-based currency to the Federal government of the United States.

Federal Reserve Notes: U.S. dollar bills; bank notes, or IOUs, issued by the Federal Reserve Bank and circulated as currency. Also, called "fiat" currency.

Federal Reserve System: The privately owned Federal Reserve Bank, which is comprised of twelve regional central banks and the Federal Reserve Board, which works with the Federal Treasury, the Social Security Administration, and the IRS.

Fiat currency: A currency that is spoken into existence by an authority figure. "Fiat" means "Let it be done."

Flash drive: A flash memory drive, thumb drive, USB flash drive, or removable Solid State Drive (SSD).

Fork: A division in a cryptocurrency blockchain, which sometimes results in the creation of a new cryptocurrency.

Fractional Reserve Lending: A fraudulent practice in which bankers pretend to lend money, which they do not have. The alleged money is created by means of double entry accounting. In the U.S. any bank may legally pretend to lend ten times more money than it claims to keep in reserve.

Holy Brotherhood, The: A term devised by the author to describe the historical enemies of witches, who continue their crimes in the present day by means of a machine of domination comprised of centralized, authoritarian systems, with their financial system at its core.

ICO (Initial Coin Offering): A method of obtaining startup capital, which is similar to a corporate Initial Public Offering (IPO), but cryptocurrency tokens are issued instead of traditional shares or stocks. Most, if not all, ICOs are pump-and-dump scams. Unlike Bitcoin and altcoins, ICOs are securities and fall under the purview of the Securities and Exchange Commission (SEC) in the U.S.

"In God We Trust" system: A term devised by the author to describe blind faith and baseless trust in authorities, whether religious or secular.

Ledger: A record of finances and transactions.

Men of God: A term devised by the author to describe those who support the Holy Brotherhood, whether through collusion or ignorance, including the faithful followers of the Sinister Trinity.

Money: A measure of value and a device by which to transfer it, which possesses some intrinsic value of its own.

Open Source Software (OSS) or Open Source Code: Computer code with a copyright license that permits anyone the rights to alter, distribute, and use it for any purpose.

Pagan: One who is not a member of the Sinister Trinity.

Peer to Peer (P2P): A system in which each individual can connect directly with another individual on a network without routing communications through a central point

QR code: "QR" stands for "Quick Response." QR codes are an example of modern steganography.

Sinister Trinity: The three world religions, Judaism, Christianity, and Islam.

Steganography: A form of cryptography involving pictures. The art and science of hiding a message in an image.

Subsystems: The smaller systems that function together to supply nourishment to the financial system, which is the heart of the Holy Brotherhood's domination machine. Examples of subsystems include those that support orthodox medicine, official science, sanctioned religion, and major media.

Trusted third party: A mediator who facilitates and arbitrates a transaction.

Wallet: A metaphor for how you store your cryptocurrency account information.

REFERENCES AND NOTES

Introduction
1. Press Release, "Three NSA Whistleblowers Back EFF's Lawsuit Over Government's Massive Spying Program," *Electronic Frontier Foundation*, July 2, 2012. https://www.eff.org/press/releases/three-nsa-whistleblowers-back-effs-lawsuit-over-governments-massive-spying-program.

Chapter 1
1. Jansen, Koos, "Audits Of US Monetary Gold Severely Lack Credibility," *Bullionstar Blogs*, March 28, 2018. https://www.bullionstar.com/blogs/koos-jansen/audits-of-us-monetary-gold-severely-lack-credibility/
2. The "Black Lodge" is described by Dion Fortune in *Psychic Self-Defense, Rider & Company, 1930.*
3. Blavatsky, Helena, *Isis Unveiled: a Master-Key to the Mysteries of Ancient and Modern Science and Theology,* Volume Two, Theology, 2nd Point Loma Edition, The Theosophical Publishing Company, Point Loma, California, 1910, p. 524. https://books.google.com/books?id=QZN7nQAACAAJ. Note: "Pleyte" refers to the Dutch Egyptologist Willem Pleyte (1836-1903).

Chapter 2
1. Kramer, Samuel Noah, *From the Tablets of Sumer: Twenty-Five Firsts in Man's Recorded History,* Indian Hills: The Falcon's Wing Press, 1956, p. 55.
2. Tacitus, Publius Cornelius, Transl. Alfred John Church and William Jackson Brodribb, Ed. Moses Hadas, *Complete Works of*

Tacitus, Random House, Inc. 1942, p. 658.

3. Ibid. p. 659.

4. Ibid.

5. Ibid.

6. Ibid.

7. Tacitus, *The Histories*, Book 5, Paragraph 9, p. 662.

8. Golan, David, "Hadrian's Decision to supplant 'Jerusalem' by 'Aelia Capitolina,'" *Historia: Zeitschrift für Alte Geschichte*, Bd. 35, H. 2 (2nd Qtr., 1986), pp. 226-239. Published by: Franz Steiner Verlag Stable URL: http://www.jstor.org/stable/4435963.

9. "The Origins of Coinage," *The British Museum.* http://www.britishmuseum.org/explore/themes/money/the_origins_of_co inage.aspx.

10. Ibid.

11. "The Hellenistic Period: The Role of the Banks," Το Σήμα του Ιδρύματος Μείζονος Ελληνισμού. www.ime.gr/chronos/06/en/economy/index308.html.

12. Heckethorn, Charles William, *The Secret Societies of All Ages and Countries: A Comprehensive Account of Upwards of One Hundred and Sixty Secret Organizations - Religious, Political and Social - from the most Remote Ages down to the Present Time, Volume 1,* George Redway, London, 1897, p. 172.

13. *Encyclopedia.com.* http://www.encyclopedia.com/books/politics-and-business-magazines/stora-kopparbergs-bergslags-ab.

14. The above dates are approximations and various authors and researchers provide slightly different dates and times for these events. Some say the Templars were never really dissolved and that many escaped persecution.

15. "De heretico comburendo," *The Oxford Companion to British History,* Oxford University Press, 2002. Retrieved June 11, 2018 from Encyclopedia.com. https://www.encyclopedia.com/history/encyclopedias-almanacs-transcripts-and-maps/de-heretico-comburendo.

16. Ibid.

17. Heckethorn, Ibid.

18. The choice of fire to torture and murder their enemies, as well as their threats of fire and brimstone, represents the destructive, elemental nature of the Sinister Trinity's god, which is a syncretic spirit of fire. He is described in as a "consuming fire" in *Deuteronomy 4:24 (KJV):* "For the Lord thy God is a *consuming fire*, even a jealous God;" *Deuteronomy 9:3:* "Understand therefore this day, that the Lord thy God is he which goeth over before thee; *as a consuming fire he shall destroy* them...;" and

Hebrews 12:29: "For our God is a consuming fire." In *Genesis 19:24*, God punishes humans by raining down fire upon them: "Then the Lord rained upon Sodom and upon Gomorrah *brimstone and fire* from the Lord out of heaven." In passages of *Exodus,* God appears as "a pillar of fire" at night as a guide, his angel appears to Moses in a "flame of fire," and his followers make "offerings made by fire," to him. *Exodus 35:3* demands "Ye shall kindle no fire throughout your habitations upon the sabbath day" likely because it was on the seventh day their fire god "rested."

19. "America's First IPO," *Museum of American Finance.* http://www.moaf.org/exhibits/americas_first_ipo/index.

20. "Brief History of IRS," *IRS*, https://www.irs.gov/uac/brief-history-of-irs. Page Last Reviewed or Updated: September 28, 2016.

21. Ibid.

22. Mullins, Eustace, *Secrets of the Federal Reserve,* Bankers Research Institute, 1983, p. 9.

23. An early twentieth-century banking insider, Congressman, and whistleblower, Louis Thomas McFadden, meticulously exposed the Federal Reserve Bank in 1934 during a speech he made in Congress, published and available in its entirety at the Internet Archive, entitled *"Congressman McFadden on the Federal Reserve Corporation,"* by Louis McFadden. It was published as: *Congressman McFadden on the Federal Reserve Corporation, Remarks in Congress, AN ASTOUNDING EXPOSURE, 1934.* https://archive.org/details/CongressmanMcfaddenOnTheFederalReserve Corporation. The inner workings of their secret banking society was widely exposed, again, by Eustace Mullins, *The World Order: Our Secret Rulers* in 1954, *The Secrets of the Federal Reserve* in 1952, and in numerous lectures available for viewing at the Internet Archive and many popular social media websites.

24. Mullins, Eustace, *Secrets of the Federal Reserve*, Bankers Research Institute, 1983, p. 101. https://archive.org/details/TheSecretsOfTheFederalReserve

25. "The Consumer Protection from Unfair Trading Regulations 2008," *Legislation.gov.uk*, No. 1277, 2008. http://www.legislation.gov.uk/uksi/2008/1277/contents/made.

26. Somoza, Anastasio, *Nicaragua Betrayed,* Western Islands, 1980.

27. Walsh, Independent Counsel, "Final Report of the Independent Counsel for Iran/Contra Matters: Volume I: Investigations and Prosecutions," *UNITED STATES COURT OF APPEALS FOR THE DISTRICT OF COLUMBIA CIRCUIT, Division for the Purpose of Appointing Independent Counsel, Division No. 86-6, Chapter 11,* August

4, 1993. https://fas.org/irp/offdocs/walsh/chap_11.htm

28. Congressman Brad Sherman [D-CA], House of Representatives, *C-Span*, October 2, 2008.

29. "Bailout Push Included Threat of Martial Law," *World Net Daily*, October 1, 2008. http://www.wnd.com/2008/10/77860.

30. "Transaction View information about a bitcoin transaction," *Blockchain.com.*
https://www.blockchain.com/btc/tx/4a5e1e4baab89f3a32518a88c31b
c87f618f76673e2cc77ab2127b7afdeda33b?show_adv=true.

31. Elliott, Francis and Gary Duncan, "Chancellor Alistair Darling on Brink of Second Bailout for Banks: Billions May Be Needed as Lending Squeeze Tightens," *The Times,* January 3, 2009.
https://www.thetimes.co.uk/article/chancellor-alistair-darling-on-brink-of-second-bailout-for-banks-n9l382mn62h.

32. Editor, "Only Three Countries Left Without a ROTHSCHILD Central Bank!" *The Event Chronicle,* January 30, 2017.
http://www.theeventchronicle.com/finanace/three-countries-left-without-rothschild-central-bank. Wars were waged on the four most recent countries to fall to the central banking system: Afghanistan, Iraq, Sudan, and Libya.

Chapter 3

1. McFadden, Ibid.
https://archive.org/details/CongressmanMcfaddenOnTheFederalReser
veCorporation

2. Press Release, "President Bush Attends Office of Faith-Based and Community Initiatives' National Conference" *The White House,* June 26, 2008.
https://georgewbush-whitehouse.archives.gov/news/releases/2008/06/20
080626-20.html.

3. *Godless Geeks*, Compiled by Atheist Coalition of San Diego,
http://www.godlessgeeks.com/LINKS/StateConstitutions2.htm.

4. Arkansas State Legislature, HJR-1009 "Amending the Arkansas Constitution to Repeal the Prohibition Against an Atheist Holding Any Office in the Civil Departments of the State of Arkansas or Testifying as a Witness in Any Court," *7th General Assembly,* February 11, 2009.
http://www.arkleg.state.ar.us/assembly/2009/R/Pages/BillInformation.as
px?measureno=HJR1009

5. Higler, M., "Kiddushin 66c," *Tractate Soferim,* New York, 1937, 15:7, p. 282.

6. "Bank Resolution and 'Bail-in' in the EU: Selected Cast Studies

Pre and Post BRRD," *World Bank Group: Finance & Markets, Financial Sector Advisory Center (FinSAC),* The World Bank Group, Vienna, Austria. http://pubdocs.worldbank.org/en/120651482806846750/FinSAC-BRRD-and-Bail-In-CaseStudies.pdf.
   7. *Malachi 3:9.*

Chapter 4
   1. (a) Du, Lisa, "Gerald Celente: My Gold Account Was Emptied By MF Global," *Business Insider,* Nov. 16, 2011. http://www.businessinsider.com/the-truth-behind-how-gerald-celente-got-screwed-by-mf-global-2011-11; (b) Du, Lisa, "Two MF Global Clients Spill About Their Frozen Accounts -- 'It Was Like Being In The Twilight Zone,'" *Business Insider,* November 9, 2011.
   2. (a) "Author Naomi Wolf Sues Bank, Says $300,000 Stolen from Accounts," *The Smoking Gun*, August 23, 2010. http://www.thesmokinggun.com/buster/naomi-wolf/author-naomi-wolf-sues-bank-says-300000-stolen-accounts; and (b) Wolf, Naomi, "Banks Siding Against the Customer in Fraud Cases," *Huffington Post, The Blog,* August 23, 2010. https://www.huffingtonpost.com/naomi-wolf/post_722_b_691188.html.
   3. "First National Bank of Montgomery v. Jerome Daly," *Justice Court, Credit River Township, Scott County, Minnesota,* December 9, 1968. https://mn.gov/law-library/assets/1968-12-09judgmentanddecree_tcm1041-115904.pdf.
   4. Salviano, Roberto, "Where the Mob Keeps Its Money," *The New York Times,* August 24, 2012. https://www.nytimes.com/2012/08/26/opinion/sunday/where-the-mob-keeps-its-money.html.
   5. Syal, Rajeev, "Drug Money Saved Banks in Global Crisis, Claims UN Advisor," *The Guardian,* December 12, 2009. https://www.theguardian.com/global/2009/dec/13/drug-money-banks-saved-un-cfief-claims.
   6. Cox, Jeff, "Big Bank have Found a New Way to Stay in the Subprime Lending business," *CNBC,* April 10, 2018. https://www.cnbc.com/2018/04/10/big-banks-have-found-a-new-way-to-stay-in-the-subprime-lending-business.html.
   7. Field, Abigail, "Justice Denied: Why Countrywide Chief Fraudster Mozilo Isn't Going to Prison," *AOL.com*, February 23, 2011. https://www.aol.com/article/2011/02/23/countrywide-mozilo-fraud-no-prison-trial-sec-mortgage-meltdown-deal-crisis/19788287.

Chapter 5

1. Das, Samburaj, "IMF Director Tells Banks: Do Not Worry about Bitcoin & Blockchain," *CCN.com*, November 5, 2015. https://www.ccn.com/imf-director-tells-banks-do-not-worry-about-bitcoin-blockchain.

2. Yuen, Stacey, "Us Bank CEOs are Likely 'Very Afraid' of Bitcoin, says Wealth Advisors," *CNBC.com*, September 20, 2017. https://www.cnbc.com/2017/09/20/us-bank-ceos-are-likely-very-afraid-of-bitcoin-says-wealth-advisor.html.

Chapter 6

1. Sweeney, John, "This World: The Mormon Candidate," Sam Collyns, Producer, James Jones, Director, *BBC Two*, Broadcasting House in Westminster, London, England, March 27, 2012. Television.

2. "Investment Portfolios Connected to the Mormon Church," *MormonLeaks.io.* https://mormonleaks.io/wiki/index.php?title=Investment_Portfolios_Connected_to_the_Mormon_Church

3. (a) Pace, Glenn L., Bishop Memorandum to Strengthening Church Members, Committee, "Ritualistic Child Abuse," July 19, 1990.; and Carlisle, Nate, "Arizona Case Shows Why Mormon Bishops are Not Reporting Sex Abuse to Police Every Time," *The Salt Lake Tribune*, May 31, 2018. https://www.sltrib.com/religion/local/2018/05/31/arizona-case-shows-why-mormon-bishops-are-not-reporting-sex-abuse-to-police-every-time-that-has-a-prosecutor-complaining-about-the-churchs-lawyers; (b) Wilkinson, Rhett, "LDS Church: Woman Speaks on Why She Yelled 'Stop Protecting Sexual Predators' at Mormon Prophets," *Inquisitr*, April 4, 2018. www.inquisitr.com/4853859/lds-church-woman-speaks-on-why-she-yelled-stop-protecting-sexual-predators-at-mormon-prophets

4. Jorisch, Avi, The Vatican Bank: The Most Secret Bank in the World, *Forbes*, June 26, 2012. https://www.forbes.com/sites/realspin/2012/06/26/the-vatican-bank-the-most-secret-bank-in-the-world/#23e92772120b.

5. Sanderson, Rachel, "The Scandal at the Vatican Bank," *Financial Times*, December 6, 2013. https://www.ft.com/content/3029390a-5c68-11e3-931e-00144feabdc0

6. Mullins, Eustace, *The World Order: A Study in the Hegemony of Parasitism,* Ezra Pound Institute of Civilization, 1985, pp. 139-198.

Chapter 7

1. Nakamoto, Satoshi, "Bitcoin: A Peer-to-Peer Electronic Cash System," *Bitcoin.org,* October 31, 2008. https://bitcoin.org/bitcoin.pdf.

2. Redman, Jamie, "Bitcoin's Relationship with 'Mark of the Beast' Theories," *Bitcoin.com,* July 27, 2017. https://news.bitcoin.com/bitcoins-relationship-with-the-mark-of-the-beast-theories.

3. (a) Chaum, David, "Blind Signatures for Untraceable Payments," *Advances in Cryptology Proceedings*, Springer-Verlag. 82 (3), 1982, pp. 199–203. http://www.hit.bme.hu/~buttyan/courses/BMEVIHIM219/2009/Chaum. BlindSigForPayment.1982.pdf; (b) Chaum, David, Amos Fiat, Moni Naor, *Proceedings on Advances in Cryptology (Santa Barbara, California, United States), Springer-Verlag,* New York, New York, 1990, pp. 319-327: http://blog.koehntopp.de/uploads/chaum_fiat_naor_ecash.pdf; and (c) Clark, Tim, "DigiCash Loses U.S. Toehold," *Cnet,* September 2, 1998. https://www.cnet.com/news/digicash-loses-u-s-toehold

4. Chaum, David. "Security Without Identification: Transaction Systems To Make Big Brother Obsolete," *Communications of the ACM* (October 1985), 1030–1044. https://link.springer.com/chapter/10.1007/3-540-39805-8_28.

5. Dai, Wei, "B-Money," *weidai.com,*1998. http://www.weidai.com/bmoney.txt.

6. Hughes, Eric. "A Cypherpunk's Manifesto," *Activism.net,* March 9, 1993. https://www.activism.net/cypherpunk/manifesto.html.

7. Ibid.

8. Wood, Robert W., "91% of IRS Seizures for 'Structuring' Involve Lawful Taxpayers," *Forbes,* April 5, 2017. https://www.forbes.com/sites/robertwood/2017/04/05/91-of-irs-seizures-for-structuring-involve-lawful-taxpayers/#4b128c355123.

9. Madore, P.H., "The Government Could Struggle to Seize Your Bitcoin," *Bitcoin Op-ed News*, May 25, 2015. https://www.cryptocoinsnews.com/government-struggle-seize-bitcoin.

10. "OHP Uses New Device To Seize Money During Traffic Stops," *News9,* June 9, 2016. http://www.news9.com/story/32168555/ohp-uses-new-device-to-seize-money-used-during-the-commission-of-a-crime.

11. *115th Congress (2017-2018),* "S.1241 - Combating Money Laundering, Terrorist Financing, and Counterfeiting Act of 2017," May 5, 25, 2017. www.congress.gov/bill/115th-congress/senate-bill/1241/text.

12. Morris, David Z., "The Equifax Hack Exposed More Data Than Previously Reported," *Fortune,* February 11, 2018. http://fortune.com/2018/02/11/equifax-hack-exposed-extra-data

13. Jim Reeds, *John Dee and the Magic Tables in the Book of Soyga*, p. 4. https://darkbooks.org/pp.php?v=1788325276.

14. Kuntz, Darcy, *The Complete Golden Dawn Cipher Manuscript*, Holmes Pub Group Llc; 1st edition, November 1, 1996, p. 33. Note: In cryptography, "polygraphy," refers to the substitution of a block of letters with a uniform secret code.

15. "The Discovery of Emil Stejnar" *A Bardon Companion*, (Excerpted from Paul Allen's Franz Bardon Research website, which is now defunct.) http://www.abardoncompanion.de/Stejnar.html.

16. Hulme, F. Edward, F.L.S., F.L.A., *Cryptography: or, The History, Principles, and Practice of Cipher-writing*, Ward, Lock and Co. Limited, London, 1841-1909. https://archive.org/details/cryptographyorhi00hulmuoft.

17. Coggine, Anthony, "Which Universities Are Offering Blockchain Courses?" *The Cointelegraph,* June 23, 2017. https://cointelegraph.com/news/which-universities-are-offering-blockchain-courses.

Chapter 8
1. *Black's Law Dictionary*, West Publishing, St. Paul, Minnesota, 1891, p. 384. http://blacks.worldfreemansociety.org/1/D/d-0384.jpg.

Chapter 9
1. Cline, Bugsy, "Stupidity vs. the Constitution... Front Royal Va Town Council Meeting 8/11/2014," *Youtube.com*, August 12, 2014. https://www.youtube.com/watch?v=JgPWpbOXdns; and (b) Cline, Bugsy, "EXCLUSIVE HD Full Meeting Video of Va. Town Battling Over Magical Arts," *Youtube.* August 24. 2014. https://www.youtube.com/watch?v=OkOdtwyT98I.

2. Pitzl-Waters, Jason, "Wiccan Denied Clergy Status in Virginia," *The Wild Hunt,* June 29, 2012. http://wildhunt.org/2012/06/wiccan-denied-clergy-status-in-virginia.html.

3. Adams, Guy, "Self-Publishers Accuse PayPal of Censorship," *Independent,* March 24, 2012. https://www.independent.co.uk/life-style/gadgets-and-tech/news/self-publishers-accuse-paypal-of-censorship-7534792.html.

4. Coker, Mark, "PayPal Revises Policies to Allow Legal Fiction," *The Official Smashwords Blog*, March 13, 2012. http://blog.smashwords.com/2012/03/paypal-revises-policies-to-allow-legal.html

5. (a) Dewey, Caitlin, "Inside the (Literal) Witch Hunt That's Devouring Etsy," *The Washington Post,* June 18, 2015.

https://www.washingtonpost.com/news/the-intersect/wp/2015/06/18/inside-the-literal-witch-hunt-thats-devouring-etsy/?utm_term=.bdf219debb99; (b) Kaufman, Sarah, "How To Get Away With Being An Etsy Witch: New Etsy witch policies are forcing metaphysical purveyors to be more legally savvy," *Vocativ,* Jun 17, 2015. http://www.vocativ.com/202676/how-to-get-away-with-being-an-etsy-witch; and (c) Etsy Services page, *Etsy*: https://www.etsy.com/help/article/4524

6. *BBC*, "Sale of Tarot Readings and Spells Banned on eBay," August 20, 2012. http://beta.bbc.com/news/technology-19323622.

7. Ward, Terence P., "The High-risk Digital World of Occult Sales and Psychic Services," *The Wild Hunt,* March 22, 2017. http://wildhunt.org/2017/03/the-high-risk-digital-world-of-occult-sales-and-psychic-services.html.

8. Pitzl-Waters, Jason, "What's Happening in Beebe, Arkansas?" *The Wild Hunt,* June 17, 2104. http://wildhunt.org/2014/06/whats-happening-in-beebe-arkansas.html and Brantley, Max, "Pagan Priest of Beebe Makes New York Times," *Arkansas Times, Arkansas Blog,* July 29, 2014.

9. "Pagan Church Zoning Issue in Beebe," *KARK.com,* Little Rock, Arkansas. https://www.kark.com/news/pagan-church-zoning-issue-in-beebe/209374647

10. "Problems in Beebe, AR," *Seeker's Temple Website.* http://www.seekerstemple.com/problems-in-beebe.

11. Ibid.

12. Mehta, Hemant, "Luciferian Church Opens Up in Texas with Christians Protesting Outside," *Patheos,* November 1, 2015. https://www.youtube.com/watch?v=SIzZYUVFdug.

http://www.patheos.com/blogs/friendlyatheist/2015/11/01/luciferian-church-opens-up-in-texas-with-christians-protesting-outside.

13. (a) Christian, Carol, "Church of Lucifer Vandalized Prior to Opening Outside Houston," *Houston Chronicle*, Friday, October 30, 2015; and (b) Greater Church of Lucifer, "Video Updated to Show Break of Window" November 2, 2015. *Facebook post.* https://www.facebook.com/Houstonsatan/videos/397457290378769.

14. Ford, Michael, "Assembly of Lightbearers Announcements and Info," *Youtube,* December 24, 2016. https://www.youtube.com/watch?v=SIzZYUVFdug.

15. Blakinger, Keri, Exorcised: Luciferian Church Looks to Start Anew After Harassment," *Houston Chronicle,* April 23, 2017. https://www.houstonchronicle.com/life/houston-belief/article/Exorcised-Luciferian-church-looks-to-start-anew-11093429.php.

16. Hawkins, John and Stephen Jones, "Real-life Witches Branded 'Terrorists' and 'Devil Worshipers' Complain They're Victims of Hate Campaign - But Police Won't Help," *Mirror,* May 26, 2017 https://www.mirror.co.uk/news/uk-news/real-life-witches-branded-terrorists-10508394.

17. Hartley-Parkinson, Richard, "'Christian Fanatics' Perform Exorcism Outside Gothic Shop and Threaten to Burn it Down," *Metro,* June 18, 2107.http://metro.co.uk/2017/06/18/christian-fanatics-perform-exorcism-outside-gothic-shop-and-threaten-to-burn-it-down-6716780.

18. Ibid.

19. Ibid.

20. "Shop Owner Says She's Target of Hate Crimes for Being a Witch, Selling Pagan Products," *CBC News*, March 1, 2017. http://www.cbc.ca/news/canada/manitoba/pagan-witch-hate-crimes-vandalism-west-end-1.4003672.

21. Ibid.

22. "Attacks on Pagan Store could Be Considered Hate Crime, Religion Prof Says," *CBC News,* March 2, 2017. http://www.cbc.ca/news/canada/manitoba/winnipeg-pagan-witch-hate-crime-1.4005972

23. Flood, Alison, US Bookstore Changes Isis Branding After Attacks," *The Guardian,* January 5, 2016. https://www.theguardian.com/books/2016/jan/05/us-bookstore-changes-isis-branding-after-attacks.

24. Erdahl, Kent, "'Isis' Bookstore in Denver Changes Sign - But Not Name - After Vandalism," *Fox 31* News, Colorado, December 29, 2015. https://kdvr.com/2015/12/29/isis-bookstore-in-denver-changes-sign-but-not-name-after-vandalism.

25. (a) Nazaryan, Alexander, "School Shooting Town Hall Attendees Screamed 'Burn Her,' NRA's Dana Loesch Claims at CPAC," *Newsweek,* February 22, 2018. http://www.newsweek.com/cpac-nra-wayne-lapierre-dana-loesch-guns-816427; and (b) Trumble, David, "Bern the Witch-Hunters," *Huffington Post Blog,* February 3, 2016. https://www.huffingtonpost.com/david-trumble/bern-the-witch-hunters_b_9145148.html.

26. Cline, Bugsy, Ibid.

27. "Witches in the 21st Century," *United Nations Human Rights, Office of the High Commissioner.* http://www.ohchr.org/EN/NEWSEVENTS/Pages/Witches21stCentury.aspx.

28. Ibid.

29. Ibid.

30. "Saudi Arabia's War on Witchcraft," *The Atlantic,* August, 2013. https://www.theatlantic.com/international/archive/2013/08/saudi-arabias-war-on-witchcraft/278701.

Chapter 10

1. Runes, Dagobert. D. , Editor, *The Diary and Sundry Observations of Thomas Alva Edison*, Philosophical Library, Inc., New York, 1948, pp. 204-244. http://itcvoices.org/wp-content/uploads/2015/03/Edison-Lost-Diary-Chapter.pdf.

2. Crookes, William, F.R.S, *Researches in the Phenomena of Spiritualism,* J. Burns, London, 1874.

3. Baines, Dan, "Crookes Residual Ectometron," *Dan Baines* website, 2016. http://www.danbaines.com/crookes-residual-ectometron.

4. Adler, Margot, *Drawing Down the Moon: Witches, Druids, Goddess Worshipers and Other Pagans in America,* Penguin Books, 2006, p. 413.

5. RxRights.org, "Why has My Online Pharmacy Stopped Taking Credit Card Payments?" *RxRights.org,* June 25, 2015. http://www.rxrights.org/why-has-my-online-pharmacy-stopped-taking-credit-card-payments

6. *Potcoin,* www.potcoin.com.

7. Parker, Luke, "Seafile Accepts Bitcoin After Paypal Shenanigans," *Brave New Coin,* July 12, 2016. https://bravenewcoin.com/news/seafile-accepts-bitcoin-after-paypal-shenanigans.

8. Ibid.

9. Ibid.

10. Ibid.

11. "How to Pay with Bitcoin and Bitcoin Cash," *Bitpay,* https://bitpay.com/pay-with-bitcoin.

12. Chokun, Jonas, "Who Accepts Bitcoins As Payment? List of Companies, Stores, Shops," *99Bitcoins,* June 19, 2016. Updated on May 28, 2017. https://99bitcoins.com/who-accepts-bitcoins-payment-companies-stores-take-bitcoins.

13. Connolly, Kate and Guy Grandjean, "Bitcoin: The Berlin Streets Where You Can Shop with Virtual Money," *UK Guardian,* April 26, 2013. https://www.theguardian.com/technology/2013/apr/26/bitcoins-gain-currency-in-berlin.

14. Suberg, William, "European Retail Giant Alza Accepts Bitcoin For Payments, Could Add Altcoins," *Cointelegraph,* May 18, 2017. https://cointelegraph.com/news/european-retail-giant-alza-accepts-bitcoin-for-payments-could-add-altcoins.

15. Helms, Kevin, "Japan's Bitpoint to Add Bitcoin Payments to

100,000+ Stores," *Bitcoin.com,* May 29, 2017.
https://news.bitcoin.com/japan-bitpoint-bitcoin-payments-stores.

16. *WikiLeaks Press Release,* December 4, 2010.
https://wikileaks.org/PayPal-freezes-WikiLeaks-donations.html.

17. Ibid.

Chapter 11

1. (a) Roberts, Daniel, "More than 75 banks are now on Ripple's blockchain network," *Yahoo Finance,* April 26, 2017.
https://finance.yahoo.com/news/75-banks-now-ripples-blockchain-network-162939601.html; and (b) Hackett, Robert, "Big Business Giants From Microsoft to J.P. Morgan Are Getting Behind Ethereum," Fortune, February 27, 2017. http://fortune.com/2017/02/28/ethereum-jpmorgan-microsoft-alliance.

2. Morris, David Z., "The Rise of Cryptocurrency Ponzi Schemes: Scammers are making big money off people who want in on the latest digital gold rush but don't understand how the technology works," *The Atlantic,* May 31, 2017.
https://www.theatlantic.com/technology/archive/2017/05/cryptocurrency-ponzi-schemes/528624.

3. Press Release, "The SEC Has an Opportunity You Won't Want to Miss: Act Now!" *U.S. Securities and Exchange Commission,* May 16, 2018. https://www.sec.gov/news/press-release/2018-88.

4. Nikilova, Maria, "CME Seeks to Intervene in Bitcoin Scam Case, Insists that Virtual Currencies are Commodities," *FinanceFeeds.com,* March 6, 2018. https://financefeeds.com/cme-seeks-intervene-bitcoin-scam-case-insists-virtual-currencies-commodities.

5. U.S. Code, Title 31, Subtitle IV, Chapter 53, Subchapter II, § 5330 - "Registration of Money Transmitting," *Cornell Law School, Legal Information Institute.* https://www.law.cornell.edu/uscode/text/31/5330.

6. Erenreich, Barbara, Dierdre English, *Witches, Midwives, and Nurses,* Feminist Press, 1972.

7. Molina, Isabel Pérez, "Knowledge and Powers," *Duoda, Women Research Center,* University of Barcelona, 2004-2008.
http://www.ub.edu/duoda/diferencia/html/en/secundario7.html.

8. Fairy, Bud, "How Marijuana Became Illegal," *Ozarkia.net.*
http://www.ozarkia.net/bill/pot/blunderof37.html.

9. McFadden, Ibid.

10. (a) Vulliamy, Ed, "How a big US bank laundered billions from Mexico's murderous drug gangs," *The Guardian,* April 2, 2011.
https://www.theguardian.com/world/2011/apr/03/us-bank-mexico-drug-gangs; and (b) Gurney, Kyra, "Funnel Accounts' Newest Money

Laundering Trend for Mexico's Cartels," *InSight Crime,* June 2, 2014. http://www.insightcrime.org/news-briefs/funnel-accounts-newest-money-laundering-trend-for-mexicos-cartels.

11. (a) Hanning, James and David Connett, "London is now the global money-laundering centre for the drug trade, says crime expert," *Independent,* July 4, 2015. http://www.independent.co.uk/news/uk/crime/london-is-now-the-global-money-laundering-centre-for-the-drug-trade-says-crime-expert-10366262.html; and (b) Tharoor, Avinash, "Banks Launder Billions of Illegal Cartel Money While Snubbing Legal Marijuana Businesses," *Huffington Post*, January 17, 2014. Updated September 22, 2014. http://www.huffingtonpost.com/avinash-tharoor/banks-cartel-money-laundering_b_4619464.html.

12. Taibbi, Matt, "Gangster Bankers: Too Big to Jail: How HSBC hooked up with drug traffickers and terrorists. And got away with it," *Rolling Stone*, February 14, 2013. https://www.rollingstone.com/politics/news/gangster-bankers-too-big-to-jail-20130214.

Chapter 12

1. O'Leary, Rachel Rose, "Crypto Kill Switch: Monro Goes to War Against Miners," *Coindesk.com,* March 26, 2018. https://www.coindesk.com/crypto-kill-switch-monero-going-war-big-miners.

2. Eastwood, Ross, "Change.org Using Monero Mining for Good but Hacking Problems Persist," *The Independent Republic,* July 24, 2018. https://theindependentrepublic.com/2018/07/24/change-org-using-monero-mining-for-good-but-hacking-problems-persist.

3. *OpenBazaar Blog,* October 19, 2017 https://www.openbazaar.org/blog/openbazaar-seller-guide-what-to-expect-in-this-decentralized-marketplace.

4. "How to Buy Bitcoins at a Bitcoin ATM," *CoinATMRadar,* October 31, 2014. https://coinatmradar.com/blog/how-to-buy-bitcoins-with-bitcoin-atm.

5. "Where can I find the private keys for my wallet?," *Coinbase*. https://support.coinbase.com/customer/portal/articles/1526452-where-can-i-find-the-private-keys-for-my-wallet.

6. Frunza, Marius-Cristian, "Solving Modern Crime in Financial Markets: Analytics and Case Studies". *Academic Press,* 2016, p. 65.

7. Hern, Alex, "Mt Gox CEO Charged with Embezzling £1.7m Worth of Bitcoin," *The Guardian,* September 14, 2015. https://www.theguardian.com/technology/2015/sep/14/bitcoin-mt-gox-

ceo-mark-karpeles-charged-embezzling.

8. De, Nikhilesh . "US Court Freezes BitConnect Assets as Lawsuits Mount," *Coindesk,* Jan 31, 2018, Updated Feb 1, 2018. https://www.coindesk.com/us-court-freezes-bitconnect-assets-as-lawsuits-mount

9. "Order Granting Temporary Restraining Order" filed January 30, 2018 in *United States District Court, Western District of Kentucky, Louisville Division*, Case No.: 3:18-cv-00058-JHM, Brian Paige, v. Bitconnect International PLC, et.al., http://bitcoinlawhub.com/resources/2018-01-30---U.S.-District-Court-W.D.-Kentucky---Temporary-Restraining-Order-Against-Bitconnect.pdf.

10. Ludwig, Sean, "Y Combinator-backed Coinbase Now Selling Over $1M Bitcoins Per Month," *Venture Beat,* February 8, 2013. https://venturebeat.com/2013/02/08/coinbase-bitcoin.

11. *Coinbase,* Ibid.

12. Feroz, Aeeshan, CEO (UK) at Coinbase, "Major Strides in Europe: Coinbase is the First Crypto Exchange to Obtain a UK Bank Account," *The Coinbase Blog,* March 14, 2018. https://blog.coinbase.com/major-strides-in-europe-coinbase-is-the-first-crypto-exchange-to-obtain-a-uk-bank-account-144f0e8ed7ce.

13. Erb, Kelly Phillips, "Coinbase Notifies Customers That It Will Turn Over Court-Ordered Data," *Forbes,* February 28, 2018.https://www.forbes.com/sites/civicnation/2018/08/01/how-tennessee-is-proving-fafsa-completion-leads-to-a-college-going-culture/#193ae3466598.

14. Case No. 3:17-cv-01431-JSC *United States District Court, Northern District of California,* "Order Re Petition to Enforce IRS Summons" Re: Dkt. Nos. 1, 37, 45. *United States, v. Coinbase, Inc. et al.,* Filed November 28, 2017. https://blockexplorer.com/news/wp-content/uploads/2017/11/365893210-US-v-Coinbase-order.pdf.

Chapter 13

1. *A Treaties on the Law of Torts or the Wrongs Which Arise Independent of Contract*, Cooley, Thomas M., LL. D., Chicago: Callaghan and Company, 1879. University of Michigan Law School website: http://repository.law.umich.edu/books/11.

2. Brandeis, Louis Dembitz and Samuel D.Warren, "The Right to Privacy," *Harvard Law Review, Vol. IV, 1890-91,* No. 5., December 15, 1890, Harvard Law Review Publishing Association, Cambridge, Mass., 1891. https://www.gutenberg.org/ebooks/37368.

BECOME A PATRON OF SOPHIA DIGREGORIO USING BITCOIN AND ALTCOINS

For more details about our Bitcoin and altcoin patronage program, please, read the short, free ebook, *The Occult Files of Sophia diGregorio Bitcoin and Altcoins Patronage Program: How to Join Our Cryptocurrency-based Patronage Program and Why We are Doing Things This Way* by Sophia diGregorio, which is available at the following websites:

Traditional Witchcraft and Occultism Wordpress Blog:
www.traditionalwitchcraftandoccultism.wordpress.com

The Occult Files of Sophia diGregorio Wordpress Blog:
www.occultfilesofsophiadigregorio.wordpress.com

Psychic Powers and Magic Spells:
www.psychic-powers-and-magic-spells.weebly.com

Winter Tempest Books at Webs:
www.wintertempestbooks.webs.com

To join our cryptocurrency-based patraonage program, please, contact one of our administrators:

Max Goddard: MaxGoddard@protonmail.com
Kipp Kelsey: KippKelsey@tutamail.com

ORDER OUR BOOKS WITH CRYPTOCURRENCY

To place an order for Winter Tempest Books using Bitcoin, Monero, Litecoin, Dogecoin, and other preferred altcoins, please, contact:

Kipp Kelsey: KippKelsey@tutamail.com

Max Goddard: MaxGoddard@protonmail.com.

MORE WINTER TEMPEST BOOKS

If you enjoyed this book, you might enjoy other Winter Tempest Books:

All Natural Dental Remedies: Herbs and Home Remedies to Heal Your Teeth & Naturally Restore Tooth Enamel by Angela Kaelin

Black Magic for Dark Times: Spells of Revenge and Protection by Angela Kaelin

The Devil's Grimoire: A System of Psychic Attack by Moribus Mortlock

Grimoire of Santa Muerte: Spells and Rituals of Most Holy Death, the Unofficial Saint of Mexico (Santa Muerte Series) (Volume 1) by Sophia diGregorio

Grimoire of Santa Muerte, Vol. 2: Altars, Meditations, Divination and Witchcraft Rituals for Devotees of Most Holy Death (Santa Muerte Series) (Volume 2) by Sophia diGregorio

How to Communicate with Spirits: Séances, Ouija Boards and Summoning by Angela Kaelin

How to Develop Advanced Psychic Abilities: Obtain Information about the Past, Present and Future Through Clairvoyance by Sophia diGregorio

How to Read the Tarot for Fun, Profit and Psychic Development for Beginners and Advanced Readers by Angela Kaelin

How to Write Your Own Spells for Any Purpose and Make Them Work by Sophia diGregorio

Magical Healing: How to Use Your Mind to Heal Yourself and Others by Angela Kaelin

Natural Remedies for Reversing Gray Hair: Nutrition and Herbs for Anti-aging and Optimum Health by Thomas W. Xander

Practical Black Magic: How to Hex and Curse Your Enemies by Sophia diGregorio

Spells for Money and Wealth by Angela Kaelin

The Traditional Witches' Book of Love Spells by Angela Kaelin

Traditional Witches' Formulary and Potion-making Guide: Recipes for Magical Oils, Powders and Other Potions by Sophia diGregorio

What's Next After Wicca? Non-Wiccan Occult Practices and Traditional Witchcraft by Sophia diGregorio

To place an order for Winter Tempest Books using Bitcoin, Monero, Litecoin, Dogecoin, and other preferred altcoins, please, contact:

Kipp Kelsey: KippKelsey@tutamail.com

Max Goddard: MaxGoddard@protonmail.com.

SOPHIA DIGREGORIO'S CRYPTOCURRENCY DONATION CODES

Donate to the author:

Monero (XMR):
41wVxhAQchuESryqAQgnyhY7Qv4McnFrFZ6Sb
9ue16AdJzmGUMuBY6zP7cZ1JBG7nVfqJRUqW
zDBhayebZwae93pNkyFnMm

Bitcoin (BTC):
196qNENpoe8DGCt8mHYcm2xZ3oKZSxxyvq

Dogecoin (DOGE):
DBNNCYZe6WWPFZokjd933bm2eHLd9gAXzy

Litecoin (LTC):
La8aBs7BPVwP87tmKH9ggL1bVuoq2x866W

Dash (DASH):
XwTZaBxDSAxnAif3SWrkMUPhGSYXet5MpG

Bitcoin Cash (BCH):
14merNRUCMBHVJUDxb3q2ktdypUxiq4Qvu

Bitcoin Gold (BTG):
GRwsULUvqeHwCJ9V14J21b1GgS25KSz6S9

Zcash (ZEC):
t1f57jkSeiTLYMC1wcEWYu8BCLmRMDnfBmp

DISCLAIMERS

www.ingramcontent.com/pod-product-compliance
Lightning Source LLC
Chambersburg PA
CBHW071737200326
41519CB00021BC/6761